DATE DUE			
OCT 17	NOV 29		
OCT 21	JAN 0 2		
OCT 9 6	OC 28 02		
35			
OCT 22	NO 19 02		
OCT 30	34/03		
NOV 24			
34 9			
OCT 3 1			
NOV 14			

530
V

VanCleave, Janice
Pratt.

Janice VanCleave's
physics for every
kid.

Janice VanCleave's

Physics for
Every Kid

The Janice VanCleave
SCIENCE FOR EVERY KID
Series

Janice VanCleave's
Physics for Every Kid

101 Easy Experiments in Motion, Heat, Light, Machines, and Sound

Janice Pratt VanCleave

John Wiley & Sons, Inc.

New York • Chichester • Brisbane • Toronto • Singapore

530
V

Illustrated by Barbara Clark

The publisher and the author have made every reasonable effort to ensure that the experiments and activities in this book are safe when conducted as instructed but assume no responsibility for any damage caused or sustained while performing the experiments or activities in *Janice VanCleave's Physics for Every Kid.* Parents, guardians, and/or teachers should supervise young readers who undertake the experiments and activities in this book.

Library of Congress Cataloging-in-Publication Data

VanCleave, Janice Pratt.
    Janice VanCleave's physics for every kid : 101 easy experiments in motion, heat, light, machines, and sound.
       p.   cm. — (Wiley science editions)
    ISBN 0-471-52505-7 (pbk.)
    ISBN 0-471-54284-9 (lib. bdg.)
    Summary: Presents 101 experiments relating to physics using materials readily available around the house.
    1. Science—Experiments—Juvenile literature. [1. Physics—Experiments. 2. Experiments.] I. Title. II. Title: Physics for every kid.
Q164.V36 1991
530′.078—dc20   90-13018

Printed in the United States of America

4 5 6 7 8 9 10-MU-959493

Printed and bound by Courier Companies, Inc.

Dedicated to my helpmate,
Wade Russell VanCleave

Preface

This is a physics experiment book, and like its predecessors *Chemistry for Every Kid*, *Biology for Every Kid,* and *Janice VanCleave's Earth Science for Every Kid,* it is designed to teach that *Science is fun!* Physics is more than a list of facts. The 101 physics experiments are designed for children ages 8 through 12. Young children will be able to successfully complete the experiments with adult supervision. Older children can easily follow the step-by-step instructions and complete the experiments with little or no adult help. Special warnings are given when adult assistance might be required.

The book contains 101 experiments relating to physics. Each experiment has a purpose, a list of materials, step-by-step instructions, illustrations, expected results, and a scientific explanation in understandable terms.

The introductory purpose for each experiment gives the reader a clue to the concept that will be introduced. The purpose is complete enough to present the goal but does not give away the mystery of the results.

Materials are needed, but in all the experiments the necessary items are easily obtained. Most of the materials are readily available around the house. A list of the necessary supplies is given for each experiment.

Detailed step-by-step instructions are given along with illustrations. Pretesting of all the activities preceded the drafting of the instructions. The experiments are *safe* and they *work*.

Expected results are described to direct the experimenter further. They provide immediate positive reinforcement to the student who has performed the experiment properly, and they help correct the student who doesn't achieve the desired results.

Another special feature of the book is the Why? section, which gives a scientific explanation for each result in terms easily understood.

This book was written to provide safe, workable physics experiments. The objective of the book is to make the learning of physics a rewarding experience and thus stimulate you to seek more knowledge about science.

Note:

The experiments and activities in this book should be performed with care and according to the instructions provided. Any person conducting a scientific experiment should read the instructions before beginning the experiment. An adult should supervise young readers who undertake the experiments and activities featured in this book. The publisher accepts no responsibility for any damage caused or sustained while performing the experiments or activities covered by this book.

Acknowledgments

I wish to express by appreciation to special friends who have helped by pretesting the activities or just giving moral support when I needed it: H. L. and Beulah Bradford, Dwayne and Rhonda Dickerson, Mark and Trica Corbett, Ruth Ethridge; Jacque Toland; Sandra Kilpatrick; David and Holly Ruiz; Nancy Rothband; Jo Childs; and Ron Ross.

Some special children helped pretest the experiments. This book was written with the hope that they will become more interested in the scientific world around them: Sterling Russell; Brandon and Erica Dickerson; Gavin and Ashley Corbett; David and Cidney Meyer; Michael, Matthew, and Krista Brantner; Jason, Caleb, and Micah Burson; Jessica, Kathryn, and Charles James; Brandon and Bryan Collins; Justin and James Hatcher; Mandy and Michelle Ethridge; and Charlcie Rothband.

A special group of helpers from Chilton High School aided in the pretesting of experiments used in this book. I am thankful to these future scientists for their help: Johnny Berry, Jr., Tracie Broughton, Loretta McClain, and Sonya Randolph.

A special note of gratitude to the members of my family who volunteer their time and sometimes unknowingly do some of the pretesting. These very special helpers are: Russell, Ginger, Kimberly, Jennifer, David, Tina, and Davin VanCleave; as well as Calvin, Ginger, and Lauren Russell; Raymond, Rachel, Ryan, Dennis, Brenda, Carol, Erin, and Amber Pratt; Patsy and Kenneth Henderson; Kenneth, Dianne, Kenneth Roy, and Robert Fleming; and Craig, Kymie, Krystie, Allen, and Megan Witcher.

My husband, Wade, receives my deepest gratitude. His love and encouragement are invaluable.

May I never forget that God is the author of science, and how much fun He has provided in allowing me to discover a small part of His wonderful creation.

Janice VanCleave

Contents

Introduction

Physics is the study of energy and matter and the relationship between them. Studying physics, like studying all sciences, is a way of solving problems and discovering why things happen the way they do. It seems that we have always tried to explain the world around us. Science began and continues due to our curiosity, and our curiosity often leads to new discoveries. The laser, for example, is a result of curious physicists wanting to know more about light. These scientists had no special purpose for the laser; they were simply investigating ways to produce powerful light beams. This book may not lead to any new scientific discoveries, but it will provide fun experiments that teach physics concepts.

We live in an exciting time for scientists. Much scientific information has already been collected and new facts are being discovered daily. Yet each new discovery makes evident that there is a storehouse of scientific knowledge waiting to be unlocked.

1

This book will help you to make the most of the exciting scientific era in which we live. It will guide you in discovering answers to questions relating to physics such as: Why is the sky blue? Why does the earth wobble? What causes a volcanic eruption? What happens during an earthquake? What causes dew to form? How can a diamond be cut so smoothly? What is inside the earth? How does rock movement produce heat? Where does rain come from and where does it go? The answers to these questions and many more will be discovered by performing the fun, safe, and workable experiments in this book.

You will be rewarded with successful experiments if you read an experiment carefully, follow each step in order, and do not substitute equipment. It is suggested that the experiments within a group be performed in order. There is some build-up of information from the first to the last, but any terms defined in a previous experiment can be found in the glossary. A goal of this book is to guide you through the steps necessary in successfully completing a science experiment and to teach you the best method of solving problems and discovering answers. The following list gives the standard pattern for each experiment in the book:

1. Purpose: This states the basic goals for the experiment.
2. Materials: A list of necessary supplies.
3. Procedure: Step-by-step instructions on how to perform the experiment.
4. Results: An explanation stating exactly what is expected to happen. This is an immediate learning tool. If the expected results are achieved, the experimenter has an immediate positive reinforcement. An error is also quickly

recognized, and the need to start over or make corrections is readily apparent.
5. Why?: An explanation of why the results were achieved is described in understandable terms. This means understandable to the reader who may not be familiar with scientific terms.

General Instructions for the Reader

1. **Read first.** Read each experiment completely before starting.
2. **Collect needed supplies.** You will experience less frustration and more fun if all the necessary materials for the experiments are ready for instant use. You lose your train of thought when you have to stop and search for supplies.
3. **Experiment.** Follow each step very carefully, never skip steps, and do not add your own. Safety is of the utmost importance, and by reading any experiment before starting, then following the instructions exactly, you can feel confident that no unexpected results will occur.
4. **Observe.** If your results are not the same as described in the experiment, carefully reread the instructions, and start over from the first step.

Measurement Substitutions

Measuring quantities described in this book are intended to be those commonly used in every kitchen. When specific amounts are given, you need to use a measuring instrument closest to the described amount. The quantities listed are not critical and a variation of very small amounts more or less will not alter the results.

The exchange between SI (metric) and English measurements will not be exact. A liter bottle can be substituted for a quart container even though there is a slight difference in their amounts. The list on page 5 is a substitution list and not an equivalent exchange.

English to SI Substitutions

English	SI (Metric)

LIQUID MEASUREMENTS

English	SI (Metric)
1 gallon	4 liters
1 quart	1 liter
1 pint	500 milliliters
1 cup (8 ounce)	250 milliliters
1 ounce	30 milliliters
1 tablespoon	15 milliliters
1 teaspoon	5 milliliters

LENGTH MEASUREMENTS

English	SI (Metric)
1 yard	1 meter
1 foot (12 inches)	⅓ meter
1 inch	2.54 centimeters
1 mile	1.61 kilometers

PRESSURE

English	SI (Metric)
14.7 pounds per square inch (PSI)	1 atmosphere

Abbreviations

atmosphere = atm
centimeter = cm
cup = c
gallon = gal.
pint = pt.
quart = qt.
ounce = oz.
tablespoon = T.
teaspoon = tsp.

liter = l
milliliter = ml
meter = m
millimeters = mm
kilometers = km
yard = yd.
foot = ft.
inch = in.

I

Electricity

1. Glow

Purpose To determine how a fluorescent tube (light bulb) works.

Materials *balloon*
               fluorescent tube

Procedure
- *Inflate and tie the balloon.*
- *Wash the outside of the fluorescent tube and thoroughly dry.*
- *In a dark room, place one end of the tube against the floor.*
- *Hold the tube upright and quickly rub the balloon up and down the outside of it.*
- *Hold the balloon near the tube.*

Results The fluorescent tube starts to glow and the light moves with the movement of the balloon. Once the tube starts glowing, even the nearness of the balloon causes light to be produced.

Why? When a fluorescent tube is connected to an electrical current, the chemicals on the tiny filaments at each end of the tube release electrons. Electrons jump from one end of the tube to the other, producing 120 flashes of light every second. Too fast to see, this ultraviolet light is invisible to the human eye. A drop of mercury inside the tube is vaporized by the electrical flashes and the vapor carries electrons to the phosphor powder coating the inside of the tube. This coating changes the ultraviolet energy into light energy that can be seen. Rubbing the balloon on the tube causes the same changes to occur, but on a smaller scale. Rubbing the balloon causes electrons to build up on the surface of the balloon. This buildup of electrons causes the mercury vapor inside the tube to become charged, and just as when the tube is connected to an electric current, the charged mercury vapor bombards the fluorescent chemicals, resulting in visible light.

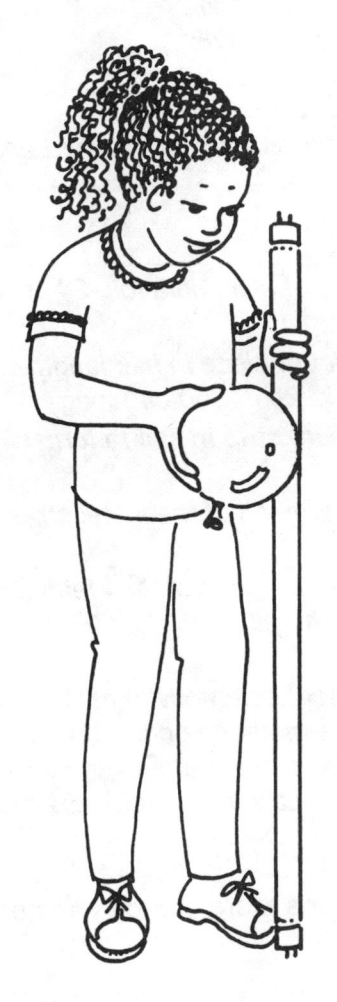

2. Conductor

Purpose To determine if all materials conduct electricity.

Materials *spring-type clothespin*
1 D-cell battery
aluminum foil
flashlight bulb
masking tape
scissors
testing materials: rubber band, paper, coins
ruler

Procedure

- *Cut a rectangle of aluminum foil, 24 in. × 12 in. (60 cm × 30 cm).*
- *Fold the aluminum piece in half lengthwise five times to form a thin strip 24 in. (60 cm) long.*
- *Cut the aluminum strip in half to form two 24-in. (30-cm) strips.*
- *Tape one end of each of the metal strips to the ends of the battery.*
- *Wrap the free end of one of the metal strips around the base of the flashlight bulb. Hold the tape in place with the clothespin.*
- *Test the electrical conductivity of the materials collected by touching the metal tip on the bottom of the flashlight bulb to one side of the material while touching the free end of the metal strip to the opposite side of the same material.*

Results The coins were the only materials that caused the bulb to glow.

10

Why? An *electric circuit* is the path through which electrons move. A *switch* is a material that acts as a bridge or pathway for the electrons. When the switch is closed, the electrons move freely, but when it is open, the electrons stop. The only materials tested that allowed electrons to flow through them were those made of metal. Touching the paper clip to one side of a piece of metal and the tip of the bulb to the other side allowed the electrons to flow out of the negative part of the battery through the aluminum tape (*conductor*) into the bulb. The electrons continue their path from the bulb through the aluminum strip and back into the positive end of the battery. As long as there is no break in the system, the electrons continue to flow and the bulb will continue to glow.

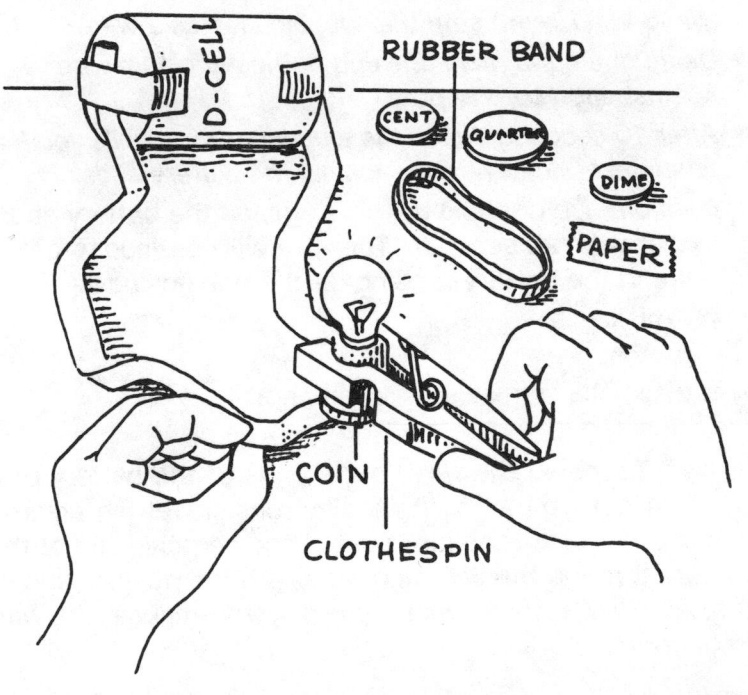

RUBBER BAND

D-CELL

CENT

QUARTER

DIME

PAPER

COIN

CLOTHESPIN

3. Hot

Purpose To discover that the flow of electrons generates heat.

Materials *1 AA battery*
aluminum foil
scissors
ruler

Procedure

- *Cut a strip from the aluminum foil, 6 in. × 1 in. (15 cm × 2.5 cm).*
- *Fold the strip of foil in half lengthwise twice to form a thin 6-in. (15-cm) strip that will be used as a wire.*
- *Using one hand, hold one end of the aluminum wire against each battery pole.*
- *After 10 seconds, touch the aluminum wire while you continue to hold the wire against the battery ends. Caution: Do not hold the wire against the battery ends longer than 20 seconds. The wire will continue to get hot and the battery is being discharged (losing its power).*

Results The aluminum wire gets hot.

Why? Touching the wire to the ends of the battery produces a path through which electrons travel (an *electric circuit*). The electrons move out of the negative end of the battery through the wire and back into the positive end of the battery. The movement of the electrons causes the wire to get hot.

When a light bulb is placed in the electrical circuit, the electrons move through the bulb. The movement of the electrons heats up the wire filament inside the bulb. The hot wire filament becomes *incandescent,* that is, it gets so hot that it gives off light.

ALUMINUM STRIP

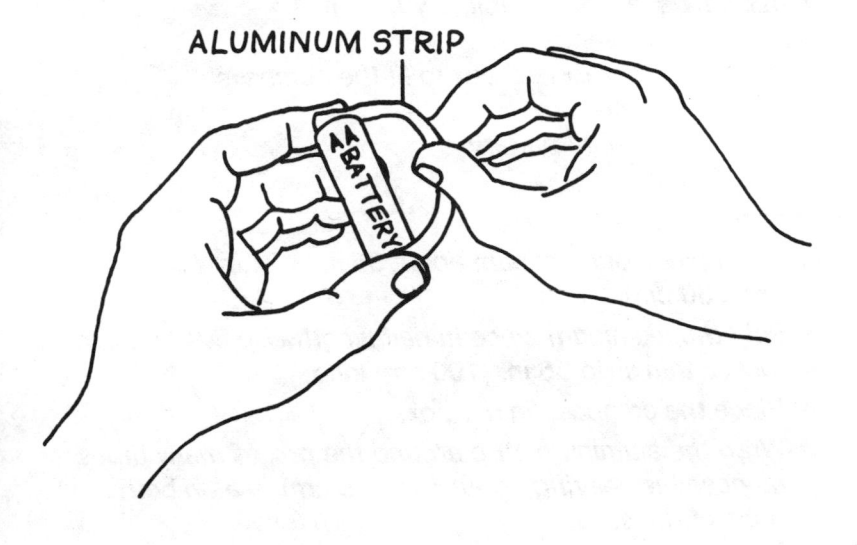

4. Galvanometer

Purpose To determine if an electric current affects a magnet.

Materials aluminum foil, 1 yd. (1 m)
compass
cardboard box to fit the compass
scissors
D-cell battery

Procedure

- Cut a piece of aluminum about 36 in. × 24 in. (100 cm × 60 cm).
- Fold the aluminum piece in half lengthwise five times to form a thin strip 36 in. (100 cm) long.
- Place the compass in the box.
- Wrap the aluminum strip around the box as many times as possible, leaving about 6 in. (15 cm) free on both ends of the strip.
- Turn the box with the compass so that the aluminum winding is pointing in a north-to-south direction.
- Tape one end of the aluminum strip to the positive end of the battery.
- Watch the compass needle as you touch the free end of the metal strip to the negative end of the battery. Alternately touch the metal strip to the battery and remove it several times.

Results The needle on the compass moves away from, and then returns to, its north-to-south direction when the metal strip is touched to and then removed from the battery.

Why? Electrons flow out of the battery, through the aluminum strip, and back to the battery. Moving electrons produce a magnetic field. Since the aluminum strip is turned in a north-to-south direction, the movement of electrons through the strip produces a magnetic field pointing east and west. The needle of the compass will be pulled toward this magnetic field, thus indicating that an electric current is flowing through the strip. The larger the current through the strip, the stronger the magnetic field that is produced.

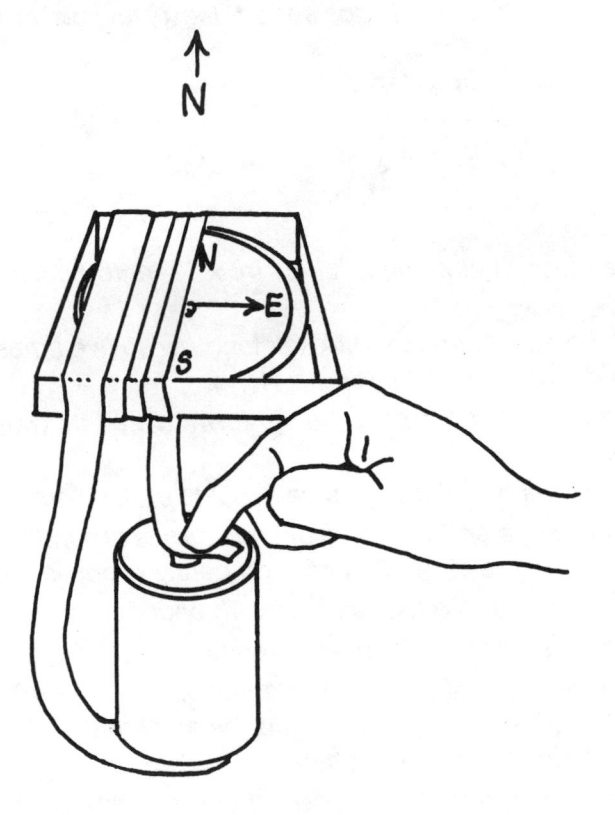

5. Potato Circuit

Purpose To determine which *battery terminal* is positive.

Materials *aluminum foil*
D-cell battery
potato
2 paper clips
2 pennies
steel wool, not soap filled (available in hard-
ware store)
masking tape
scissors
1 push tack

Procedure

- *Cut a piece of aluminum foil 24 in. × 12 in. (60 cm × 30 cm).*
- *Fold the aluminum piece in half lengthwise five times to form a thin strip 24 in. (60 cm) long.*
- *Cut the aluminum strip in half to form two 24-in. (30-cm) strips.*
- *Rub the pennies with the steel wool to clean them.*
- *Wrap the free ends of the aluminum strips around the pennies. Leave about half of each penny exposed.*
- Secure the tape to the pennies with paper clips.
- Ask an adult to cut the potato in half.
- *Insert the edge of the pennies about 1/2 inch (1 cm) apart into the sliced part of the potato, being careful to keep the aluminum strips attached.*
- *Use the tack to mark the position of the penny connected to the positive end of the battery.*

16

■ *Remove the pennies after one hour.*
■ *Examine the holes made by the pennies.*

Results The hole around the penny connected to the metal tape leading to the positive terminal is green.

Why? Connecting the penny to the positive pole of the battery gives the copper in the penny a positive charge. When the positively charged copper particles combine with negative particles in the potato, a green copper compound is formed. You can use this experiment to identify the positive terminal of any battery.

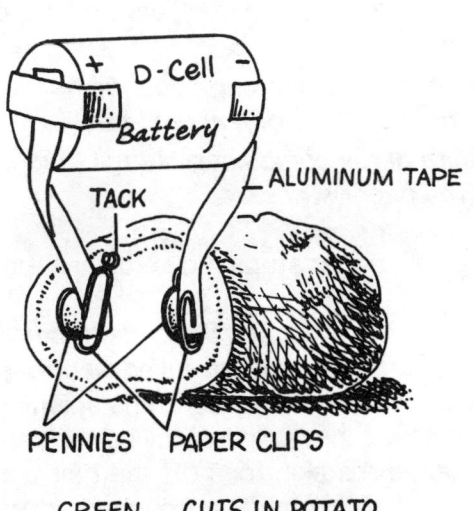

6. Streamers

Purpose To charge an object with static electricity.

Materials *comb*
tissue paper
scissors
ruler

Procedure
- *Cut a strip of tissue paper about 3 in. × 10 in. (7.5 cm × 25 cm).*
- *Cut long, thin strips in the paper, leaving one end uncut (see diagram).*
- *Quickly move the comb through your hair several times. Your hair must be clean, dry, and oil-free.*
- *Hold the teeth of the comb near, but not touching, the cut end of the paper strips.*

Results The thin paper strips move toward the comb.

Why? Static means stationary. *Static electricity* is the build-up of negative charges, which are called *electrons*. Matter is made up of atoms, which have electrons spinning around a positive center called the *nucleus*. Moving the comb through your hair actually rubs electrons off the hair and onto the comb. The side of the comb that touched your hair has a build-up of electrons, making that side negatively charged.

The paper strip is made of atoms. Holding the negatively charged comb close to the paper causes the positive part of the atoms in the paper to be attracted to the comb. This attraction between negative and positive charges is strong enough to lift individual strands of paper.

18

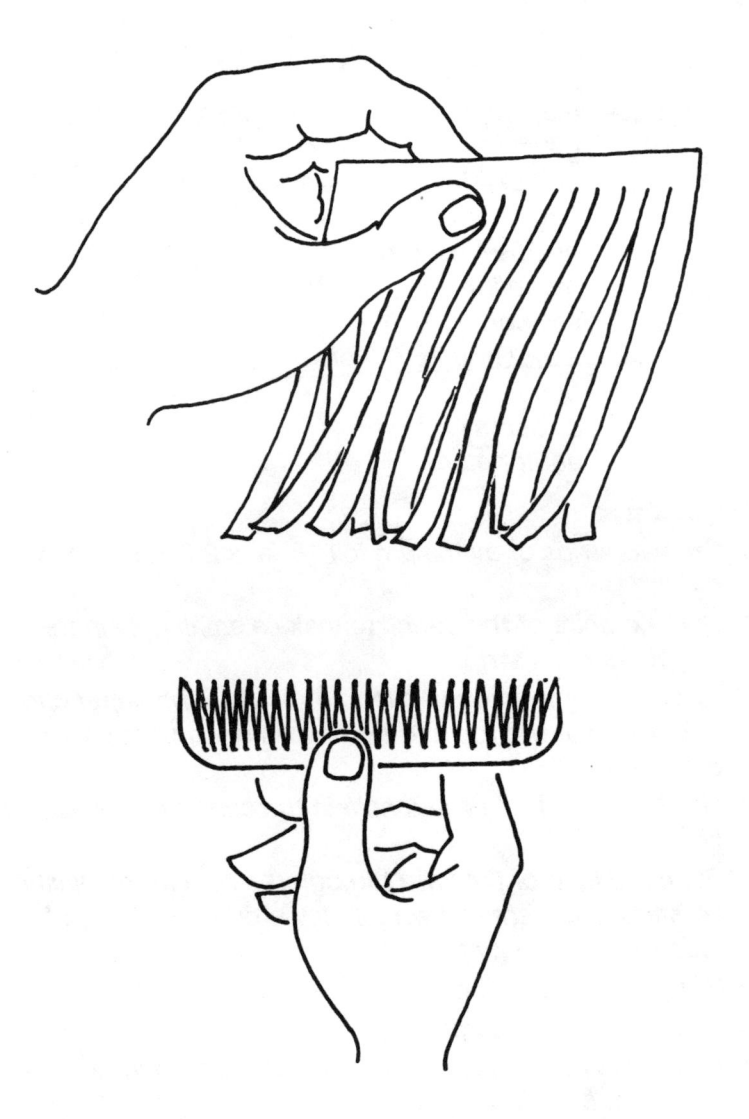

7. Move Away

Purpose To demonstrate that electrons move.

Materials aluminum foil, as thin as possible
glass jar, 1 qt. (1 liter)
soft plastic lid to cover top of jar
large paper clip
needle-nosed pliers
modeling clay
scissors
balloon, 9 in. (23 cm)
pencil
scissors
adult helper

Procedure

- Cut two strips of aluminum foil 1/2 in. × 2 in. (1 cm × 5 cm).
- Use the point of the pencil to make a small hole in the top of each foil strip.
- Ask an adult to use the pliers to reshape the paper clip into a loop at the top and two hooks on the bottom (see diagram).
- Use the pencil to make a hole in the center of the plastic lid.
- Push the loop of the wire through the hole in the plastic lid and mold a small piece of clay around the base of the loop to hold it in place.
- Hang the foil strips on the wire's hooks.
- Place the lid on top of the jar.
- Inflate the balloon and rub it on your hair. Your hair must be clean, dry, and oil-free.
- Hold the balloon near the loop, on top of the jar.

20

Results The metal leaves move apart when the charged balloon is held near the metal loop.

Why? The instrument made is called an *electroscope,* or charge detector. The metal strips move when the loop is placed in an electrically charged area. *Electrons* are rubbed off of your hair and onto the balloon, charging the surface of the balloon with a negative charge. Holding the negatively charged balloon near the metal loop causes the negative electrons on the surface of the metal to move away from the balloon because like charges *repel*, that is, move away from each other. Electrons move down the metal wire and build up on the aluminum strips. Since like charges repel each other, the two negatively charged strips move apart.

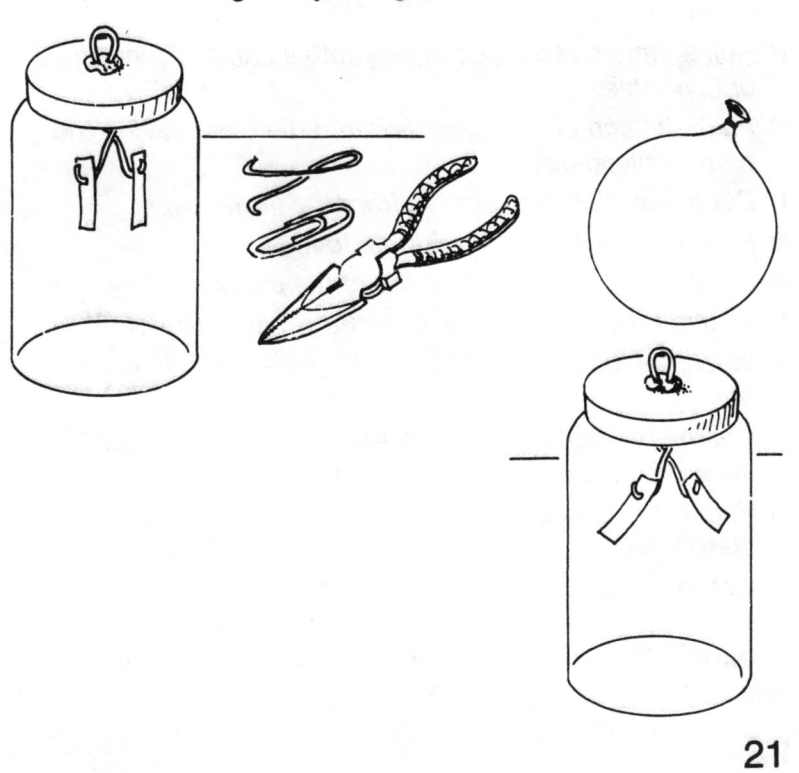

8. Don't Touch

Purpose To demonstrate the force of attraction between charged particles.

Materials *modeling clay*
push tack
tissue paper
scissors
clear plastic glass
balloon, small enough to hold in your hand
ruler

Procedure

- *Roll a dime-sized piece of clay into a ball and press it onto a table.*
- *Push the end of the push tack into the clay, leaving the point sticking up.*
- *Cut a 1-in. (2.5-cm) square from the tissue paper.*
- *Fold the paper in half to form a tent.*
- *Balance the paper tent on top of the pin point.*
- *Position the plastic glass over the pin and paper tent.*
- *Inflate the balloon to a size that is easily held in your hand.*
- *Charge the balloon by rubbing it on your hair about 10 times. Your hair must be clean, dry, and oil-free.*
- *Hold the charged balloon near, but not touching, the plastic glass.*
- *Watch the paper tent for any movement.*

Results The paper tent turns and falls off of the tip of the tack.

22

Why? The paper tent, balloon, and hair are all examples of matter, and all matter is made of atoms. Atoms have positively charged protons in their center (the *nucleus*) and negatively charged electrons spinning around the nucleus. The balloon becomes negatively charged on the side that is rubbed on the hair because the electrons are rubbed off the hair and collect on the balloon. The negatively charged balloon attracts the positive part of the paper tent. This attraction is strong enough to pull the paper off the tack.

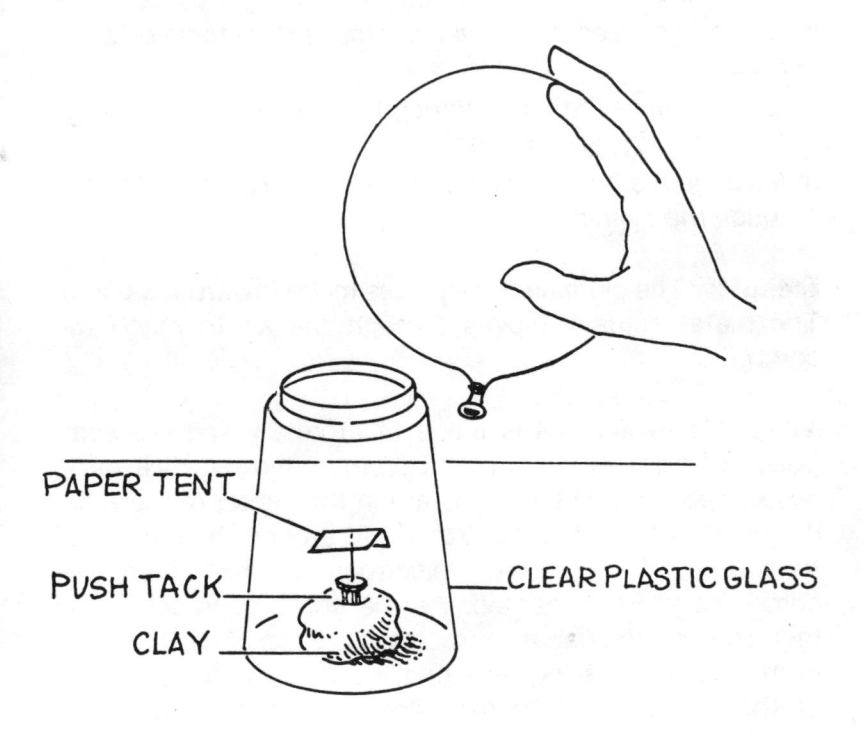

PAPER TENT

PUSH TACK

CLAY

CLEAR PLASTIC GLASS

9. Tinkle

Purpose To demonstrate the effect of static electricity.

Materials *comb*
aluminum foil
scissors

Procedure
- *Cut ten tiny pieces of aluminum foil and lay them on a table.*
- *Quickly move the comb through your hair. Your hair must be clean, dry, and oil-free.*
- *Hold the teeth of the comb above the foil pieces. Do not touch the aluminum.*

Results The aluminum foil pieces move toward the comb. The metal actually moves through the air to reach the comb.

Why? Aluminum foil is made of atoms, which are composed of positive parts called *protons* and negative parts called *electrons*. The protons are in the center of the atom (the *nucleus*) and the electrons spin around the outside of the nucleus. The comb rubs electrons off your hair and becomes negatively charged. As the comb approaches the metal pieces, the negative electrons in the metal move away from the comb, leaving more positive charges on the surface of the metal. Like charges *repel* each other and unlike charges attract. The attraction between the negatively charged comb and the positive area on the metal is strong enough to overcome the downward pull of gravity, and the metal pieces move through the air to stick to the comb.

10. Snap

Purpose To demonstrate how *static* charges produce sound.

Materials *large paper clip*
piece of wool: a scarf, coat, or sweater made of
100% wool will work
clear plastic sheet
scissors
modeling clay

Procedure

- *Cut a plastic strip about 1 in. × 8 in. (2.5 cm × 20 cm).*
- *Use the clay to stand the paper clip upright on a table.*
- *Wrap the wool around the plastic strip and quickly pull the plastic through the cloth. Do this quickly at least three times.*
- *Immediately hold the plastic near the top of the paper clip.*

Results A snapping sound can be heard.

Why? Electrons are rubbed off the wool and onto the plastic. The electrons clump together until the addition of their energy is great enough to move them across the span of air between the plastic and the metal clip. The movement of the electrons through the air produces sound waves, resulting in the snapping sound heard.

11. Close Encounter

Purpose To demonstrate attracting and repelling forces between objects due to their electrical charges.

Materials 2 round balloons, 9 in. (23 cm)
masking tape
string, 2 yd. (2 m)
clean, dry, oil-free hair
marking pen

Procedure

- *Inflate both balloons and tie their ends. Use the marking pen to label one balloon A and the other B.*
- *Cut the string in half and attach one piece to the end of each balloon.*
- *Tape the free ends of the strings to the top of a door frame so that the balloons hang about 8 in. (20 cm) apart.*
- *Stroke balloon A on your hair about 10 times and gently release it. What happens?*
- *Stroke one of the balloons on your hair about 10 times and hold it while an assistant rubs the second balloon on your hair. Gently release both balloons. Now what happens?*

Results The two balloons are attracted to each other when only one balloon is stroked against hair and repel each other when both balloons are rubbed against hair.

Why? Matter is made up of atoms, which have negatively charged electrons spinning around a positive nucleus. Electrons are rubbed off the hair and collect on balloon A; thus the balloon becomes negatively charged. Since like charges

28

repel each other, these negative charges on balloon A repel the electrons in the atoms of balloon B, causing B's surface to be more positively charged. Each balloon now has a different charge, so they are attracted to each other.

Rubbing both balloons on the hair results in a build-up of negatively charged electrons on their surfaces. The balloons repel each other because they have the same charge. There are enough repulsive and attractive forces between the balloons to cause them to move without being touched.

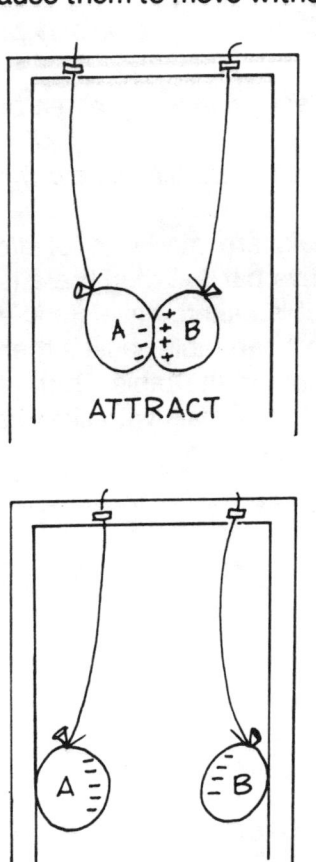

ATTRACT

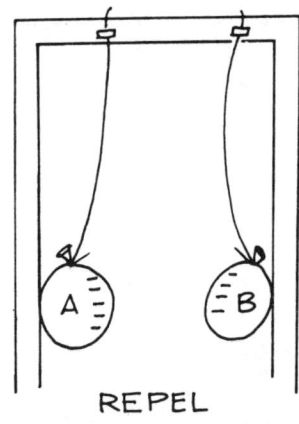

REPEL

12. Sticky?

Purpose To demonstrate the effect of static electricity.

Materials *cellophane tape*

Procedure

- *Press two pieces of tape onto a table, leaving a small piece hanging over the edge.*
- *Hold the ends of the tape and quickly pull both pieces off of the table.*
- *Bring the two pieces near each other, but not touching.*

Results The pieces of tape move away from each other.

Why? All materials are made up of *atoms,* which have positive and negative parts. When there is a gain or loss of charges, the object is said to have *static electricity.* Pulling the tape pieces from the table causes them to pick up electrons from the atoms in the table. Both pieces of tape are negatively charged. Materials with like charges *repel* (push away from) each other.

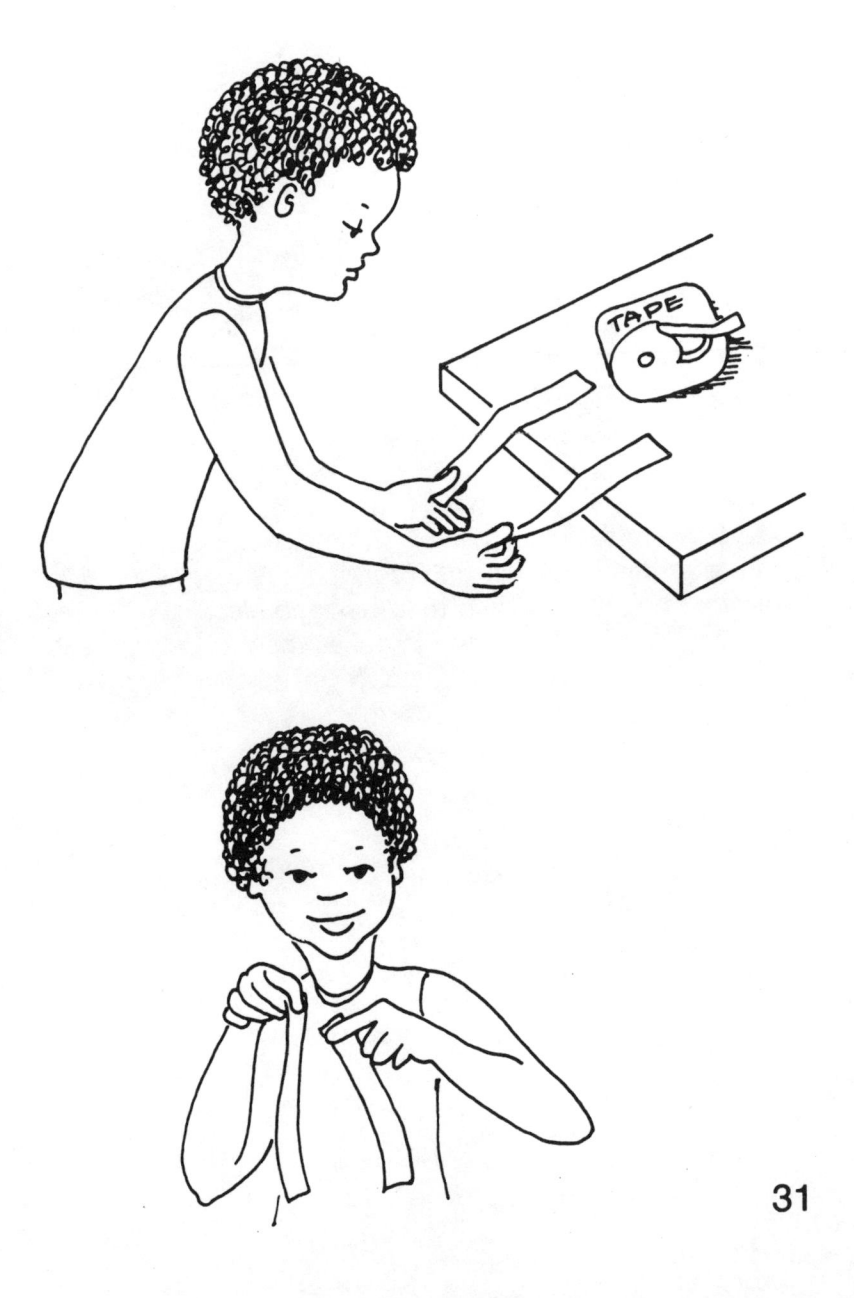

31

II
Magnets

13. Right Hand

Purpose To demonstrate how to find the magnetic north pole.

Materials *apple or tennis ball*

Procedure
- *Hold the apple in your right hand, and assume it is the earth, spinning on its axis.*
- *Wrap your fingers around the sides of the apple, with your thumb sticking up.*

Results Your fingers point in the same direction that the earth spins. Your thumb points in the same direction as does the earth's magnetic north pole.

Why? The entire earth, including the liquid core, spins in a counterclockwise direction. Electrons are rubbed from this turning liquid, but they continue to move in the same direction as the earth. A magnetic field is always present around moving electrons. The direction of the magnetic field can always be determined by using your right hand. Point your fingers in the direction of the moving electrons, and your right thumb will indicate the direction of the magnetic north pole.

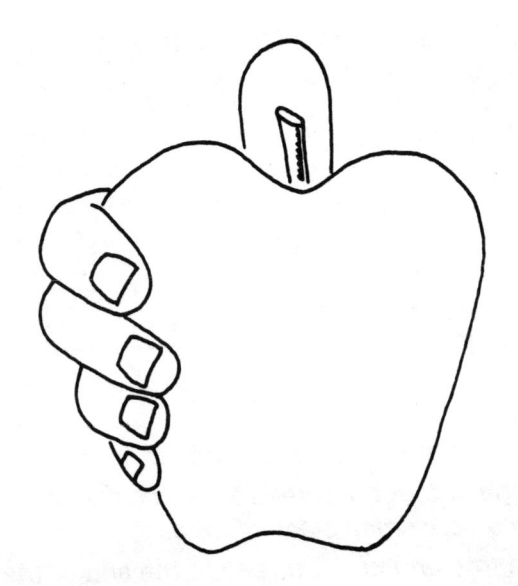

14. Swinger

Purpose To determine if the north end of a magnet always points to the earth's magnetic north pole.

Materials compass
sewing thread
small paper clip
cellophane tape
ruler
magnet
book

Procedure

- Cut a 12-in. (30-cm) piece of thread.
- Attach one end of the thread to the center of the paper clip with a very small piece of tape.
- Tape the free end of the thread to the end of the ruler.
- Set the book on the edge of a table and insert the free end of the ruler between the book's pages so that the ruler hangs over the edge of the table.
- Place the paper clip on the magnet.
- Remove the clip from the magnet and allow it to swing freely.
- Observe the direction toward which the end of the paper clip points.
- Use the compass to determine what direction that is.
- Move the ruler to different positions and observe the direction the paper clip points each time.

Results One end of the paper clip points south and the other points north. Moving the ruler does not affect the direction that the paper clip points.

Why? The earth behaves as if a large bar magnet were inside it causing magnetic materials to be attracted to its opposite ends. The north end of this imaginary magnet produces the earth's north magnetic pole, and the north ends of all magnets are attracted to this pole. The north ends of magnets are really north-seeking poles. Placing the paper clip on the magnet causes the atoms inside the paper clip to line up in a north-to-south direction. One end of the paper clip will continue to point toward the earth's magnetic north as long as the atoms inside are lined up. Moving the position of the ruler or string will not affect the direction of the free-hanging magnetized paper clip.

Note: The paper clip will remain magnetized for only a short period. You may need to put the paper clip on the magnet to realign the electrons.

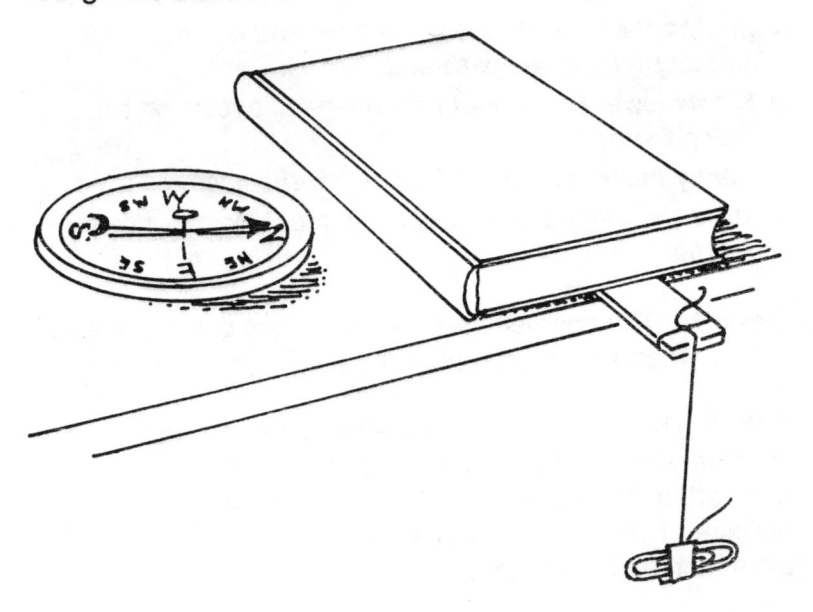

15. Floating Needle

Purpose To use a magnetic force to move a floating object.

Materials *glass bowl, 2 qt. (2 liter)*
sewing needle
sewing thread
masking tape
bar magnet

Procedure

- *Fill the bowl three-quarters full with water.*
- *Cut two 12-in. (30-cm) pieces of thread.*
- *Tape both pieces of thread to one side of the bowl, about 1 in. (2.5 cm) apart.*
- *Stretch the thread across the bowl and lay the needle across both pieces of thread.*
- *Slowly lower the thread until the needle rests on the water's surface.*
- *Gently move the thread from under the needle.*
- *Move the magnet near, but never touching, the floating needle.*

Results The needle floats on the surface of the water and moves when the magnet moves.

Why? The surface of the water acts like a thin skin. This imaginary skin is due to the attraction of water molecules to each other. The needle is able to float and move across the surface of the water in response to the magnetic force of attraction by the magnet.

16. Suspended Airplane

Purpose To use magnetic force to suspend a paper airplane.

Materials *steel straight pin*
sewing thread, 12 in. (30 cm)
tissue paper
bar magnet
scissors

Procedure

- *Cut a small wing about 1 in. (2.5 cm) long from the paper.*
- *Insert the pin through the center of the paper wing to make an "airplane."*
- *Tie the thread to the head of the pin.*
- *Place the magnet on the edge of a table with the end of the magnet extending over the edge of the table.*
- *Place the airplane on the end of the magnet.*
- *Slowly pull on the string until the airplane is suspended in the air.*

Results The airplane remains airborne as long as it stays close to the magnet.

Why? The strength of attraction between two magnets depends on how orderly the *magnetic domains* (clusters of atoms that behave like tiny atoms) are in the magnets. The atoms in the pin are randomly arranged before the pin touches the magnet. The number of atoms that arrange themselves into clusters (domains) and line up in the pin when it is placed on the magnet depends on the strength of the magnet.

The pin and magnet both have magnetic properties. They pull on each other with enough force to overcome the downward pull of gravity, which allows the airplane to remain suspended.

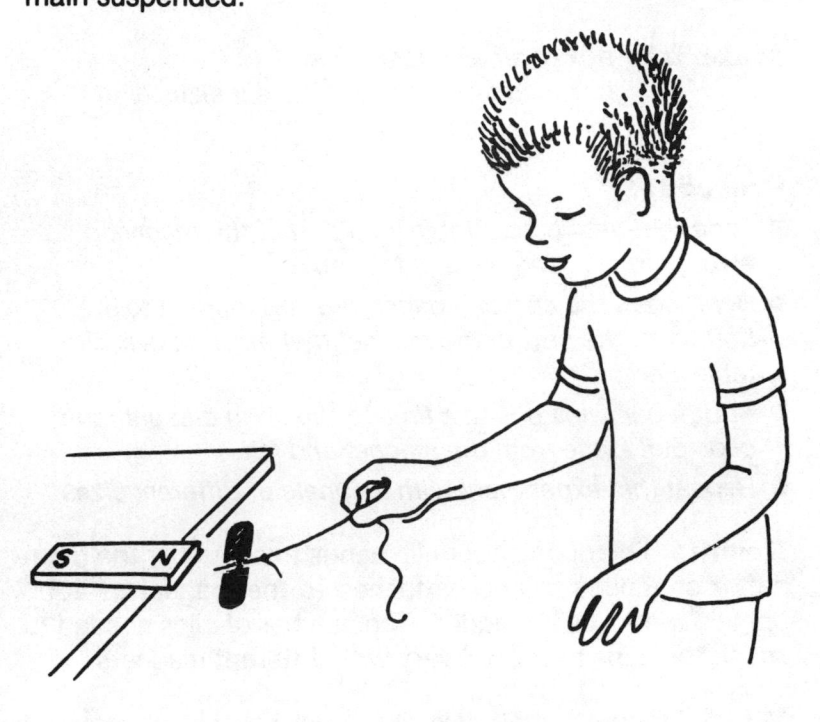

17. Magnetic Strength

Purpose To determine the strength of a magnetic force field.

Materials *box of small paper clips*
bar magnets, several different sizes
masking tape

Procedure

- *Tape the magnet to a table with part of the magnet extending over the edge of the table.*
- *Bend open the end of a paper clip and touch it to the bottom of the part of the magnet that extends over the table's edge.*
- *Add paper clips one at a time to the open clip until the clips pull loose from the magnet and fall.*
- *Repeat this experiment with magnets of different sizes.*

Results The open paper clip hangs freely under the magnet. It continues to hang attached to the magnet as additional paper clips are added. The number of clips needed to cause the clips to fall will vary with different magnets.

Why? Magnetic materials like steel contain clusters of atoms that behave like tiny magnets. These atomic clusters are called *magnetic domains.* Holding the paper clips near a magnet causes the magnetic domains inside to line up. The strength of the magnet increases as the number of domains pointing in the same direction increases. A weak magnet has a weak magnetic field around it, so its effect on magnetic materials such as paper clips is small. The number of paper clips that your magnet is able to support depends on its magnetic strength.

42

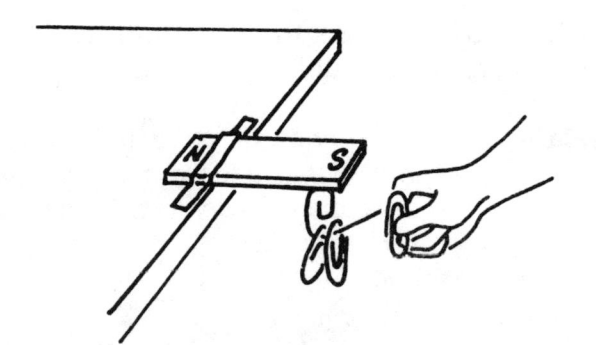

18. Force Field

Purpose To demonstrate the pattern of magnetic force fields around magnets of different shapes.

Materials *magnets, a variety —round, bar, U-shaped*
iron filings: Remove the iron filings from a
magnetic disguise set found at a toy store.
notebook paper
paper cup

Procedure

- *Pour the iron filings into the paper cup.*
- *Place the magnets on a table.*
- *Cover the magnets with a sheet of paper.*
- *Sprinkle a thin layer of iron filings on the paper over the magnets.*
- *Observe the iron filing patterns.*

Results The iron filings form a pattern of lines around the magnets. The long magnet has a buildup of filings around both ends.

Why? A *magnetic field* is the area around a magnet in which the force of the magnet affects the movement of metal objects. The iron filings are pulled toward the magnets when they enter the magnetic field. The magnetic force increases as the filings near the magnet. The force field has equal strength around the round magnet, but the force fields at the ends of rectangular magnets are always stronger than the force fields in the middle of the magnets.

19. Magnetic Shake-up

Purpose To show the effect that shaking has on a magnet.

Materials *iron filings from Experiment 18*
compass
magnet
drinking straw
modeling clay

Procedure

- *Fill the straw three-quarters full with iron filings.*
- *Use the clay to seal both ends of the straw.*
- *Lay the iron-filled straw on the magnet for one minute.*
- *Being careful not to shake the straw, pick it up and move it near the compass.*
- *Observe any movement of the compass needle.*
- *Shake the straw several times, and again hold the straw near the compass.*
- *Observe any movement of the compass needle.*

Results The compass needle is attracted to the straw before the straw is shaken, but there is no movement of the compass needle after the straw is shaken.

Why? Magnetic materials have magnetic domains (clusters of atoms that act like tiny magnets) that point in the same direction. The iron filings become magnetized when brought near the magnet because the magnetic domains in the iron filings line up. The compass needle is attracted to the magnetized, iron-filled straw. Shaking the straw rearranges the pieces of iron, causing the magnetic domains in the iron to be randomly arranged, and thus the iron filings lose their magnetic properties.

46

20. Electromagnet

Purpose To demonstrate that an electric current produces a magnetic field.

Materials *wire, 18-gauge, insulated, 1 yd. (1 m)*
6-volt battery
long iron nail
paper clips

Procedure

- *Wrap the wire tightly around the nail, leaving about 6 in. (15 cm) of free wire on each end. (This wire-wrapped nail will be used in other experiments.)*
- *Have an adult strip the insulation off both ends of the wire.*
- *Secure one end of the wire to one pole of the battery.*
- *Touch the free end of the wire to the other battery pole while touching the nail to a pile of paper clips.*
- *Lift the nail while keeping the ends of the wire on the battery poles.*
- *When the nail starts to feel warm, disconnect the wire end you are holding against the battery pole.*

Results The paper clips stick to the iron nail.

Why? There is a magnetic field around all wires carrying an electric current. Straight wires have a weak magnetic field around them. The strength of the magnetic field around the wire was increased by coiling the wire into a smaller space, placing a magnetic material—the nail—inside the coil of wire, and increasing the electrical flow through the wire—attaching a battery. The iron nail became magnetized and attracted the paper clips.

48

21. Line Up

Purpose To demonstrate how electricity and magnetism are related.

Materials *wire-wrapped nail from Experiment 20*
6-volt battery
masking tape
cardboard, 6 in. (15 cm) square
iron filings from Experiment 18
scissors

Procedure

- *Ask an adult to punch a hole through the center of the cardboard, using a nail.*
- *Insert the wire-wrapped nail through the hole in the cardboard.*
- *Make the cardboard sit flat by placing it on the roll of tape.*
- *Attach one end of the wire to either battery pole.*
- *Sprinkle a thin layer of iron filings on the cardboard around the coiled wire.*
- *Attach the free wire to the open battery terminal.*
- *Observe the pattern made by the iron filings.*
- *Disconnect the wire.*

   Caution: The nail and wires will get hot if left connected to the battery. Be sure to interrupt the circuit by disconnecting one wire from one pole.

Results The iron filings form a starburst pattern around the coil of wire.

Why? There is a magnetic field around all wires carrying an electric current. The magnetic effect around the wire can be increased by coiling more wire into a smaller space, increasing the electrical flow through the wire, and placing the iron nail in the coil of wire. The iron filings are pulled toward the magnetized nail and form a starburst pattern around the coil of wire.

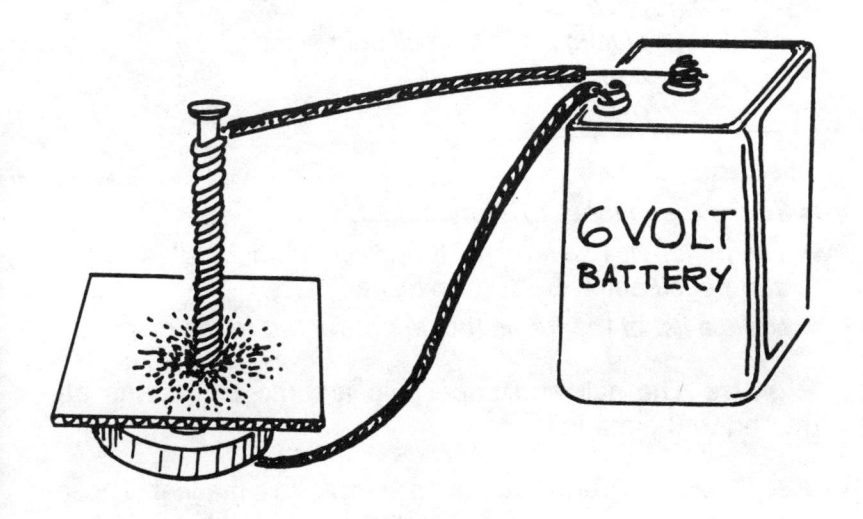

22. Attractive?

Purpose To determine what is attracted to a magnet.

Materials *magnet*
coins: penny, nickel, quarter, dime
iron nail
paper clip
aluminum foil, small coin-sized piece
pencil
paper

Procedure

- *Touch the magnet to every object.*
- *The pencil has several parts: eraser, metal band, wood, and pencil point. Be sure to test all parts.*
- *Make a list of the items that are attracted to the magnet.*

Results The nail and paper clip are the only items attracted to the magnet.

Why? Most materials are not attracted to a magnet. Those that are can become magnetized themselves. Magnetic materials contain *magnetic domains,* clusters of atoms that behave like tiny magnets. When the domains are randomly arranged, the material is not magnetic, but if the domains are all lined up, the material has magnetic properties. The steel paper clip and iron nail contain magnetic domains that were in random order before being touched by the magnet. The magnet's magnetic force pulled on the magnetic domains in the nail and paper clip, causing them to point toward the magnet. The lining up of the domains magnetized the materials. The magnet and the now-magnetized paper clip and nail are attracted to each other.

53

NON-MAGNETIC

MAGNETIC DOMAINS

MAGNETIC

23. Keeper

Purpose To determine how metals affect a magnetic field.

Materials *aluminum foil*
steel spatula
bar magnet
4 small paper clips

Procedure

- *Lay the paper clips on a table and cover them with a sheet of aluminum foil.*
- *Set the magnet on the foil over the clips.*
- *Raise the magnet and observe any movement of the clips.*
- *Position the clips so that they lay under the spatula.*
- *Set the magnet on top of the spatula.*
- *Lift the spatula with the magnet and observe any movement of the clips.*

Results The magnet attracts the paper clips through the aluminum foil. The magnet does not attract the paper clips through the steel spatula.

Why? The magnetic force field passes through the aluminum, but the steel blade restricts the movement of the force field. The steel blade is attracted to the magnet, but the metal provides another path for the magnetic field. This new path is through and around the steel blade. The steel keeps the lines of force closer to the magnetic field, acting as a barrier to other magnetic materials.

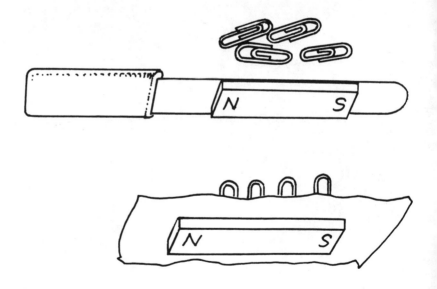

III
Buoyancy

24. Rising Bottle

Purpose To demonstrate how salt affects buoyancy.

Materials *large-mouthed jar, 1 gal. (4 liter)*
small bottle with a lid
table salt, 2 cups
measuring cup, 1 cup (250 ml)
large stirring spoon

Procedure

■ *Fill the jar three-quarters full of water.*

■ *Carefully place the closed small bottle in the water. It should float on the surface. If it does not, get a smaller bottle.*

■ *Remove the small bottle and add a small amount of water.*

■ *Close the lid and replace the bottle in the gallon jar of water. The bottle should slowly sink to the bottom. Remove or add water to the small bottle until it sinks at a slow rate when placed in the water.*

■ *Remove the bottle and add ½ cup of salt to the large jar of water. Stir.*

■ *Replace the small bottle in the jar and observe its position in the water.*

■ *Continue to add ½ cup of salt at a time until 2 cups have been added. Observe the position of the small bottle in the salty water after each measure of salt has been added.*

Results The bottle rises in the water as more salt is added.

58

Why? Adding water to the small bottle makes it heavy enough to sink. The sinking bottle pushes water out of its way. The weight of the water pushed aside equals the amount of upward force on the bottle. This force is called the *buoyancy force*. A small amount of salt increases the buoyancy force, but it is not enough to overcome the downward force resulting from the weight of the bottle and water inside. The larger amount of salt increases the buoyancy force, allowing the bottle to float on the surface of the heavy salty water.

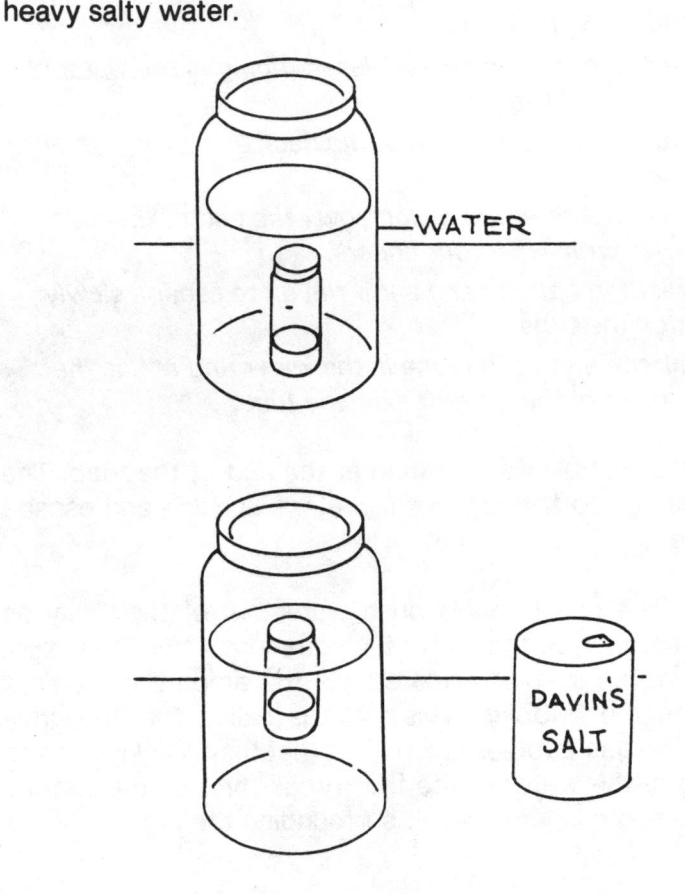

25. Bubbler

Purpose To determine why bubbles rise in liquids.

Materials *large-mouthed jar, 1 gal. (4 liter)*
clear plastic tubing, 20 in. (50 cm)
small balloon

Procedure

- *Fill the jar with water.*
- *Place one end of the clear plastic tubing in the water at the bottom of the jar.*
- *Inflate the balloon and twist the neck to prevent the air from escaping.*
- *Slip the mouth of the balloon over the end of the tube. Hold securely with your fingers.*
- *Untwist the balloon and allow the air to escape slowly through the tube.*
- *Watch the end of the tube in the water and notice the movement of the air as it exits the tube.*

Results Bubbles are formed at the end of the tube. The bubbles rise to the top of the water's surface and escape into the air.

Why? The gas bubbles push water out of their way as they emerge from the end of the aquarium tube. The weight of the water pushed aside equals the amount of upward force on the bubbles. This force is called the *buoyancy force*. The gas bubbles are so light that they quickly push to the top of the water where they break through the water's surface and mix with the air surrounding the jug.

60

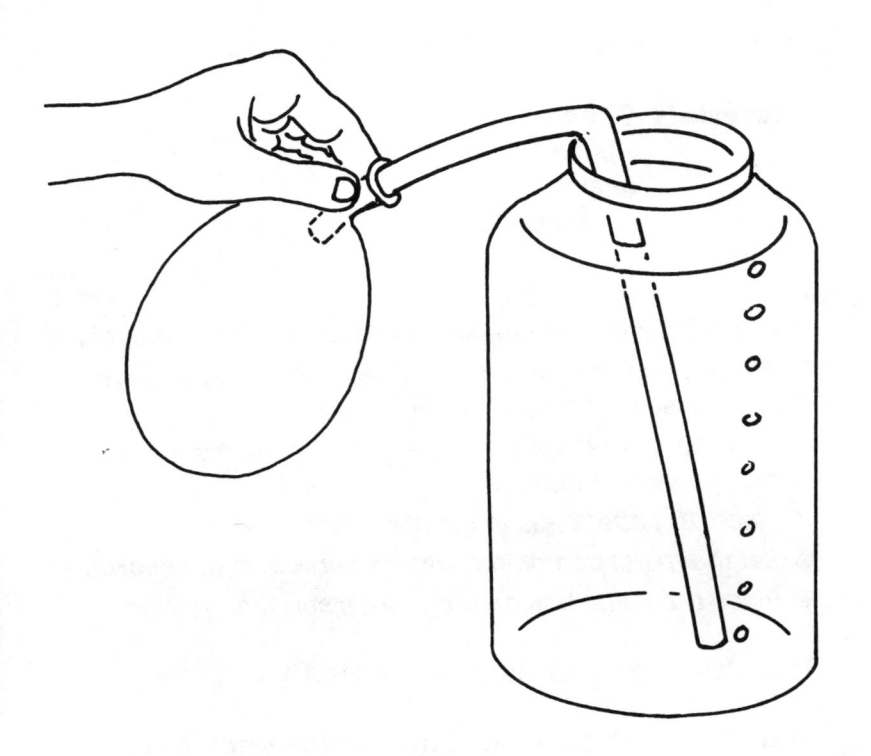

26. Floating Boat

Purpose To determine how a heavy ship floats.

Materials *20 paper clips*
aluminum foil
ruler
bucket of water

Procedure

- *Cut two 12-in. (30-cm) squares from the aluminum foil.*
- *Wrap one of the metal squares around 10 paper clips and squeeze the foil into a tight ball.*
- *Fold the four edges of the second aluminum square up to make a small square pan.*
- *Place 10 paper clips in the metal pan.*
- *Set the metal pan on the water's surface in the bucket.*
- *Place the metal ball on the water's surface.*

Results The metal pan floats and the ball sinks.

Why? The ball and pan both have the same weight, but the ball takes up a smaller space than does the pan. The amount of water pushed aside by an object equals the force of water pushing upward on the object. The ball pushes less water out of its way than does the larger pan, and thus there is not enough upward force to cause it to float. Large ships are very heavy, but they have hollow compartments filled with air, which increases their buoyancy.

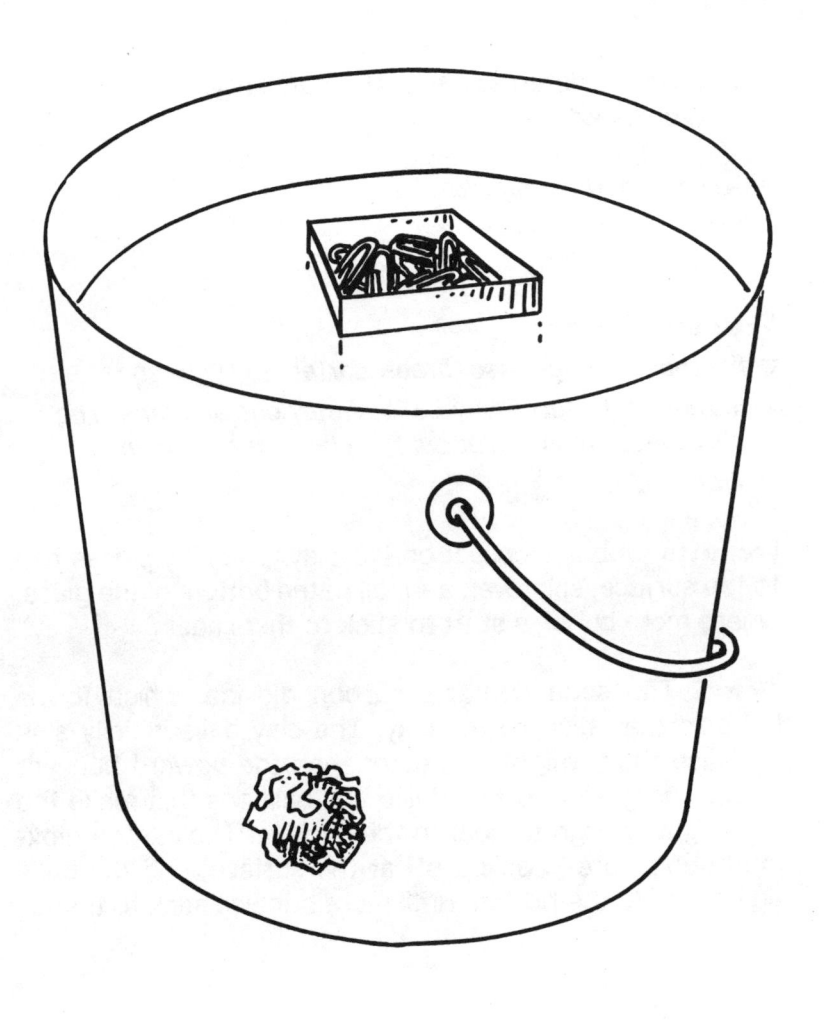

27. Risers

Purpose To determine how the buoyancy of a substance can be changed.

Materials *drinking glass*
club soda
modeling clay

Procedure
- *Fill the drinking glass three-quarters full with soda.*
- *Immediately add 5 tiny balls of clay one at a time. The clay pieces must be about the size of a rice grain.*
- *Wait and watch.*

Results Bubbles collect on the clay. The clay pieces rise to the surface, spin over, and fall to the bottom of the glass, where more bubbles start to stick to them again.

Why? The soda contains carbon dioxide, which forms bubbles that stick to the clay. The clay balls initially sink because their weight is greater than the upward buoyant force. The gas bubbles act like tiny balloons that make the balls light enough to float to the surface. The carbon dioxide bubbles are knocked off at the surface, and the balls again sink to the bottom until more bubbles stick to them.

CLAY

IV
Gravity

28. Up Hill

Purpose To determine the effect that an object's center of gravity has on motion.

Materials *2 yardsticks (meter sticks)*
3 books, each at least 1 in. (2.5 cm) thick
masking tape
2 funnels of equal size

Procedure
- *Put two books 30 in. (90 cm) apart on the floor.*
- *Place the remaining book on top of one of the other books.*
- *Position the yardsticks) on top of the books to form a V shape with the open part of the letter on the double book stack.*
- *Tape the bowls of the funnels together.*
- *Place the joined funnels at the bottom of the track formed by the yardsticks.*

Results The funnels roll up the hill.

Why? The funnels are not defying the laws of gravity. Actually, as the joined funnels move, their *center of gravity* (the point where the weight is equally distributed) moves downward. Notice that the center of the joined funnels gets closer to the floor as it moves along the raised yardsticks.

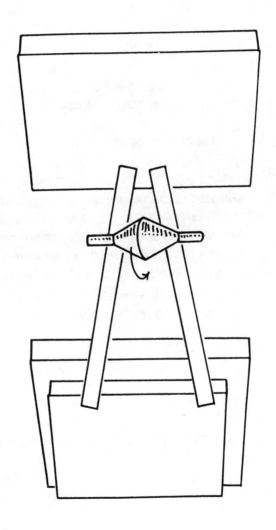

29. Bigger

Purpose To determine if size affects the speed of a falling parachute.

Materials *small plastic garbage bag*
string
2 small washers
scissors
ruler

Procedure

- *Cut eight separate strings about 20 in. (50 cm) long.*
- *Cut a 12-in. (30-cm) square from the plastic.*
- *Tie a string to each corner of the plastic square.*
- *Tie the four free ends of the strings together in a knot. Be sure the strings are all the same length.*
- *Use a string about 4 in. (10 cm) long to attach the washer to the knot in the parachute strings.*
- *Make a larger parachute using a 24-in. (60-cm) square of plastic and the four remaining strings.*
- *Attach the washer to the parachute with a 4-in. (10-cm) piece of string as before.*
- *To test the parachutes, hold each in the center of the plastic sheet. Flatten the plastic.*
- *Fold the plastic in half.*
- *Loosely wrap the string around the folded plastic.*
- *Throw the parachutes up into the air one at a time, and observe the time it takes for each to reach the ground.*

Results The larger parachute opens and floats to the ground more slowly than does the smaller parachute.

70

Why? Objects push against air as the force of *gravity* causes them to fall. This upward push of the air molecules is called *air resistance*. Objects with a large surface, such as the large parachute, have more air resistance. If the object has a large surface and a small weight, the upward push of air can equal the downward force due to weight, causing the object to gently float downward like a feather. Some insects have such a small weight as compared to their surface area that instead of falling from a height, they float down and therefore are not harmed.

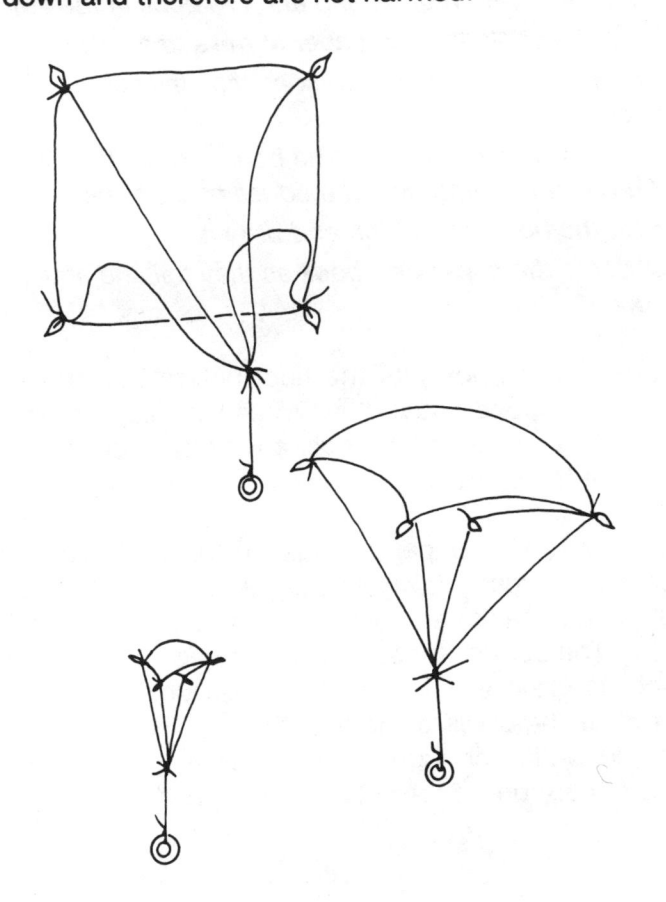

30. Same Speed

Purpose To demonstrate that gravity pulls all things down at the same speed.

Materials *paper*
book larger than the paper

Procedure
- *Hold the paper in one hand and the book in the other.*
- *Drop the book and the paper at the same time.*
- *Observe the paper and book as they fall and strike the floor.*
- *Place the paper on top of the book. Do not have any of the paper hanging over the edges of the book.*
- *Hold the book waist high and drop it.*
- *Observe the paper and book as they fall and strike the floor.*

Results The book hits the floor before the paper when they are dropped separately. When the paper is on top of the book, it falls with the book, and both—book and paper—land at the same time.

Why? When the paper is lying on the book, the objects fall together because *gravity* equally pulls on both objects. Air pushing up against an object slows the speed of the falling object. The book's speed is not changed much because its weight is great enough to produce a downward force to overcome the upward force of the air. The paper by itself does not weigh enough to overcome the upward force of the air; therefore, it falls at a slower speed.

31. Pendulum

Purpose To determine if weight affects the speed of a pendulum's swing.

Materials *watch*
            outside playground swing
            ruler
            2 helpers

Procedure

- *Hold the seat of swing and move back 3 or 4 steps.*
- *Have your helper place the ruler on the ground in front of your feet.*
- *Ask your helper to start timing when you release the swing and to report the end of 10 seconds.*
  Caution: Do not push the swing, just release it.
- *Count the number of swings back and forth in 10 seconds.*
- *Ask one of your helpers to sit in the swing.*
- *Pull the swing back until your feet are behind the ruler as before.*
- *Again have your helper start the timer when you release the swing and report the end of 10 seconds.*
  Caution: Do not push the swing.
- *Count the number of back-and-forth swings in 10 seconds.*

Results The number of back-and-forth swings is the same with and without a person sitting in the swing.

Why? The number of back-and-forth swings was the same regardless of the weight placed on the swing because

74

gravity pulled on the swing, causing it to fall when released. The speed during the swings changed, but the changes were the same for each weight that was swung. The speed increased as the swing or pendulum approached the vertical position and slowed as it moved upward where it stopped. Pendulums stop at the highest position of their swing before beginning the downward swing (that is due to the pull of gravity). The speed of this back-and-forth movement stays the same for each weight because the beginning height of the pendulum's swinging path remained the same each time and the downward pull on gravity is the same on all substances regardless of their weight.

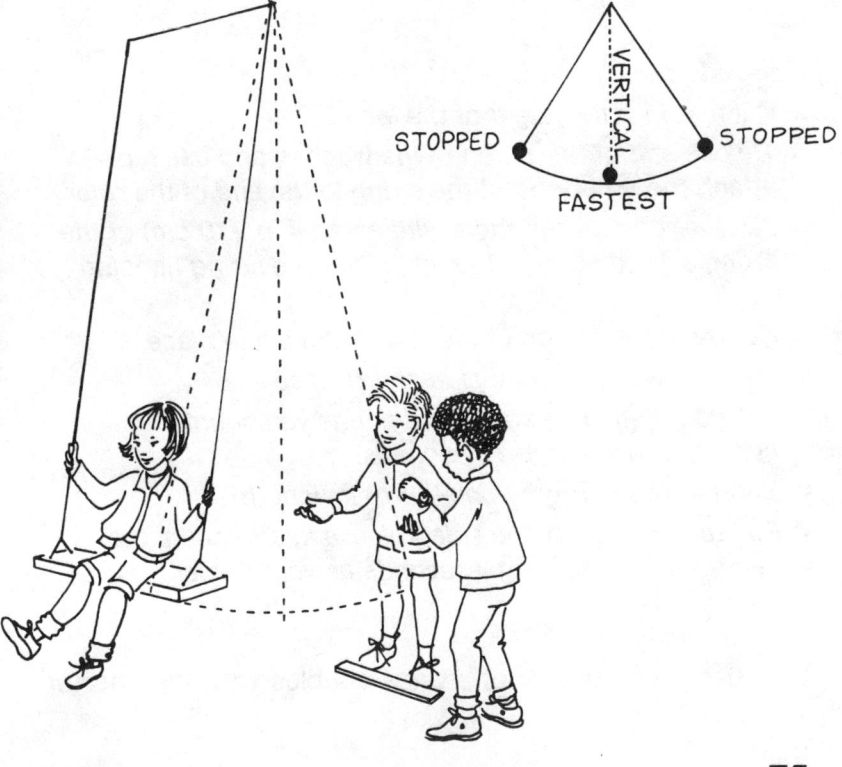

STOPPED · VERTICAL · STOPPED

FASTEST

32. Timer

Purpose To determine how the length of a pendulum affects the time of each swing.

Materials *string*
washer
scissors
ruler
heavy book
table
stop watch or watch with a second hand
helper

Procedure

- *Cut a string the height of the table.*
- *Tie one end of the string to the washer, and use tape to attach the other end of the string to the end of the ruler.*
- *Lay the ruler on the table with about 4 in. (10 cm) of the string extending over the edge, and the string hanging freely.*
- *Lay the book on top of the ruler to hold it in place.*
- *Pull the washer to one side and release it.*
- *Ask your helper to start the timer as you count the number of swings in 10 seconds.*
- *Shorten the string by one-fourth its length.*
- *Pull the washer to one side, release it, and count the number of swings in 10 seconds as your helper records the time.*

Results The number of swings doubles with the shorter string.

Why? Galileo has been credited for discovering the relationship between the length of a pendulum and the time of its swing. The story told is that he observed the swinging of a great lamp while in church and timed the swings by comparing it with his pulse beat. He later discovered that the time of a swing depends on the length of the pendulum and that the time decreases by one-half if the string is one-fourth the original length.

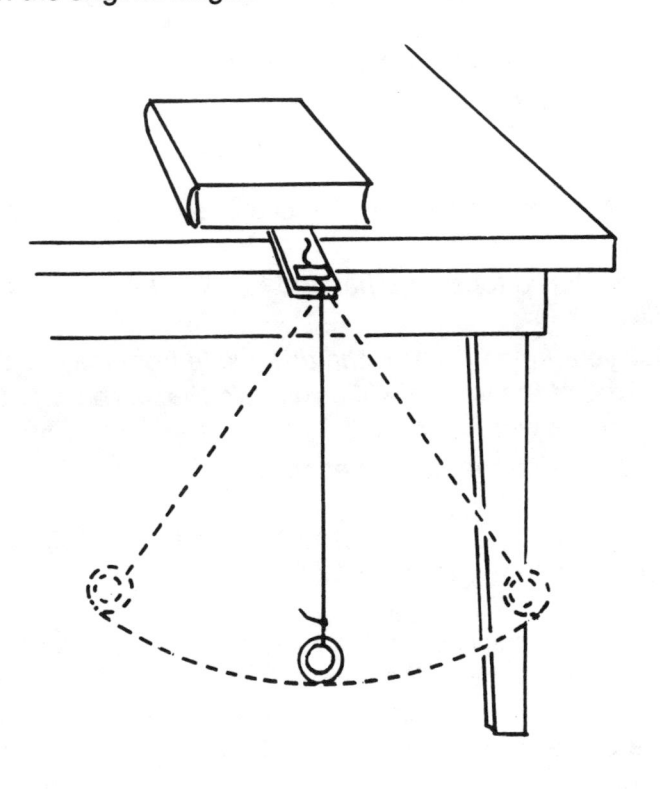

33. Shake Up

Purpose To determine how shape affects speed.

Materials *card table*
2 books of equal thickness
1 large roll of masking tape
2 jar lids, the same size
marble
masking table
helper

Procedure

■ *Tilt the card table by placing one book under two of the legs.*

■ *Place the lid tops together and tape their edges to form a disk.*

■ *Ask your helper to hold the disk made from the lids at the top of the incline while you hold the marble and tape roll in line with the disk.*

■ *Release all three objects at once.*

Results The marble rolls fastest, with the lid disk coming in second, and the tape roll last.

Why? The rolling speed is related to the distribution of weight around the object's *center of gravity*. This distribution is called the *moment of inertia*. The closer the weight is to the center of gravity, the smaller the moment of inertia and the faster the object can rotate. The center of gravity of all the objects in this experiment is at their geometric center, but each has a different weight distribution. The hollow tape roll's weight is located farthest away from the

78

center of gravity. It has the largest moment of inertia and the slowest rotation speed. The marble's weight is closest to its center of gravity, resulting in it having the fastest rotating speed.

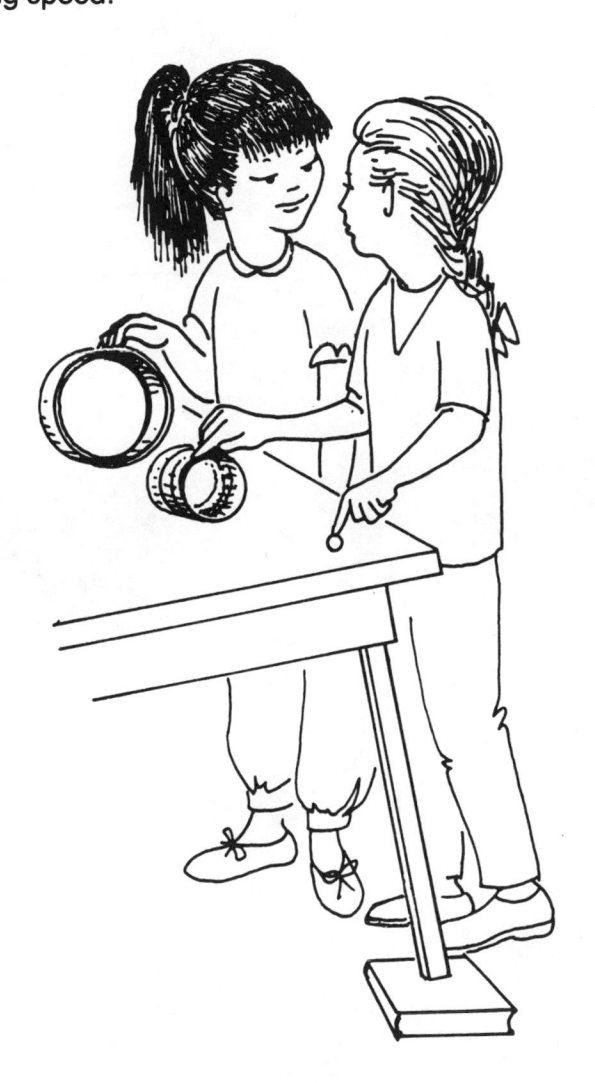

V
Balance

34. Over the Edge

Purpose To demonstrate that the center of gravity is the balancing point of an object.

Materials *yardstick (meter stick)*
hammer (wooden handled hammer works best)
string, 12 in. (30 cm)

Procedure
- *Hold the ends of the string together and tie a knot about 2 in. (5 cm) from the ends.*
- *Insert the hammer and yardstick through the loop.*
- *Position the end of the yardstick on a table's edge.*
- *The handle of the hammer must touch the yardstick and the head of the hammer will extend under the table.*
- *Change the position of the hammer until the whole unit—yardstick, string, and hammer—balances.*

Results The unit balances with only a small amount of the yardstick touching the table.

Why? The hammer, string, and yardstick all act as a single unit with a *center of gravity*. The center of gravity is the point where any object balances. The forces pushing down around this balancing point are all equal. The dashed line in the diagram allows you to visualize the center of gravity. The heavy hammer head counterbalances the weight on the left side of the balancing point.

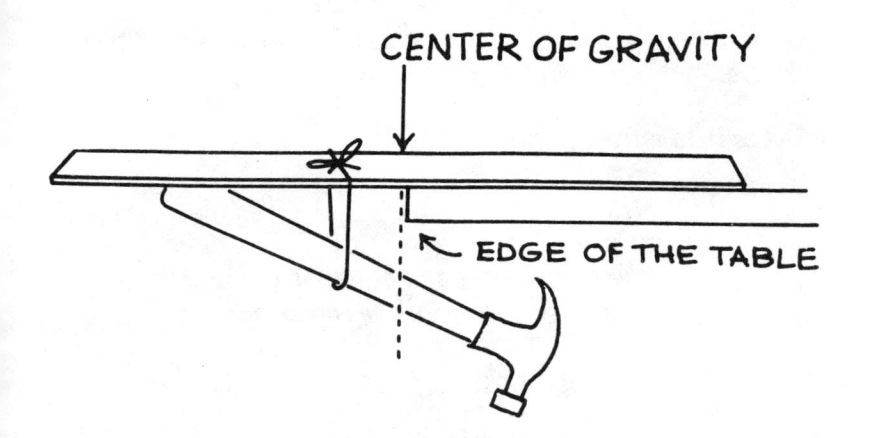

CENTER OF GRAVITY

EDGE OF THE TABLE

35. Straw Balance

Purpose To determine how center of gravity affects a balance.

Materials straight pin
small index card
straw
scissors
2 wooden blocks of equal height and not as wide as the length of the straw
ruler
marking pen

Procedure

■ Use the ruler to find the center of the straw and mark the spot with the marking pen.

■ Cut a 1-in. (2.5-cm) slit in each end of the straw, as shown in the diagram. The slits should be in the same relative position on each end.

■ Divide the index card in half by folding and cutting along the fold.

■ Insert the paper pieces in the slits on each end of the straw to form two flat surfaces that are parallel with each other. These papers will act as weighing pans.

■ Punch the straight pin through the center of the straw, leaving an equal amount of the pin sticking out on each side of the straw.

■ Position the two wooden blocks on a table and place the ends of the pin on the edges of the blocks.

■ Move the paper pieces outward to increase the downward pressure on the weighing pan and inward to decrease the pressure.

84

- *Position the paper weighing pans to make the straw level with the table.*

Results Moving the paper weighing pans in and out causes the straw to rotate around the pin. A position was found to make the straw level with the table.

Why? The position of the paper weighing pans affects the balance of the straw. The farther the paper is moved away from the pin, the more downward is the rotation of the straw on that side. Children of different weights experience the same thing when sitting on a see-saw. They move away from the center of the board to increase their downward pressure and closer to the center to decrease it. The straw is balanced when the position of the paper places the *center of gravity* of the balance at the place where the pin is inserted. Center of gravity is the point where an object balances. The downward pressure on both sides of the center of gravity point is always the same.

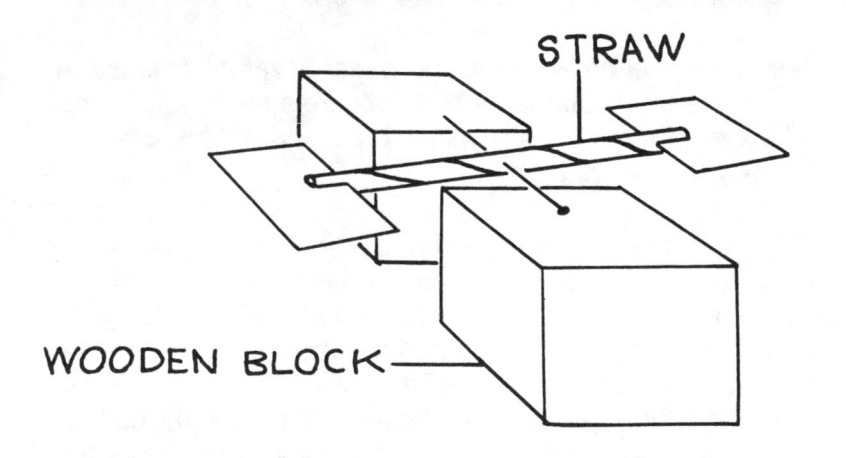

STRAW

WOODEN BLOCK

36. Creature Weigh-In

Purpose To compare the weight of a paper creature with that of paper punches.

Materials *balance from Experiment 35*
paper hole puncher
index card
scissors
pencil

Procedure

- *Draw your version of a space creature on half of the index card.*
- *Cut out the creature and place it on one of the balance's paper weighing pans.*
- *Cut paper punches from the remaining portion of the index card and continue to place them on the empty paper weighing until the straw is level with the table.*

Results The end holding the paper creature falls down, but starts to rise as paper punches are added to the opposite weighing pan. Too many punches lift the creature above the balancing point.

Why? The downward pull that *gravity* has on an object is called its *weight*. Placing the paper creature on one side of the balance increases the weight on that side. Adding the paper punches on the opposite pan begins to balance the weight of the creature. When the total weight of the paper punches equals the weight of the paper creature, the balance will be level with the table. The level balance indicates that the pull of gravity is the same on both sides of the balance.

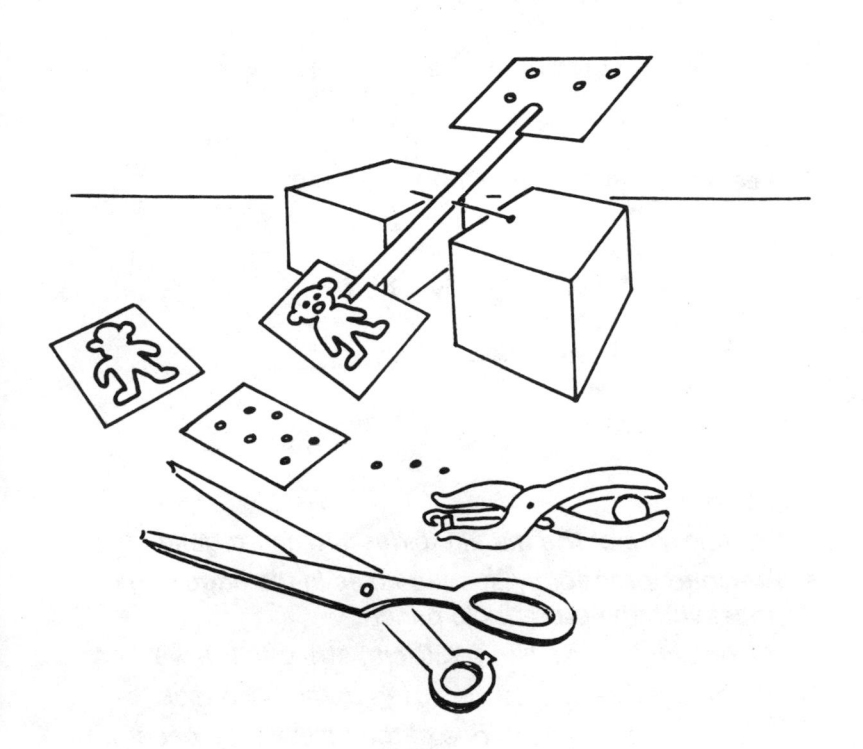

37. Downward

Purpose To determine where an object's center of gravity is.

Materials *paper hole punch*
push tack
washer
string, 12 in. (30 cm)
scissors
manila folder
ruler
tackboard

Procedure

- *Cut one side of the manila folder into an irregular shape.*
- *Punch four randomly spaced holes in the edge of the paper with the paper hole punch.*
- *Tie one end of the 12-in. (30-cm) string to the washer.*
- *Attach the free end of the string to the push tack.*
- *Stick the tack through one of the holes in the paper and into the tackboard.*
- *Allow the paper and string to swing freely.*
- *Use the ruler to mark a line on the paper next to the string.*
- *Move the tack to the other holes and mark the position of the hanging string each time. Do this for all four holes.*
- *Place the paper on the end of your index finger. Your finger is to be below the point where the lines cross.*

Results The paper balances on your finger.

88

Why? The earth's gravitational force—*gravity*—pulls downward on everything. *Center of gravity* is the place on an object that acts as if all the gravitational force pulling on the object is pulling from that spot. The center of gravity point of the paper is that point where the four lines cross. The paper can be balanced at this spot because all the paper's weight is evenly spaced around its center of gravity. Hold your finger under that point and observe the balance.

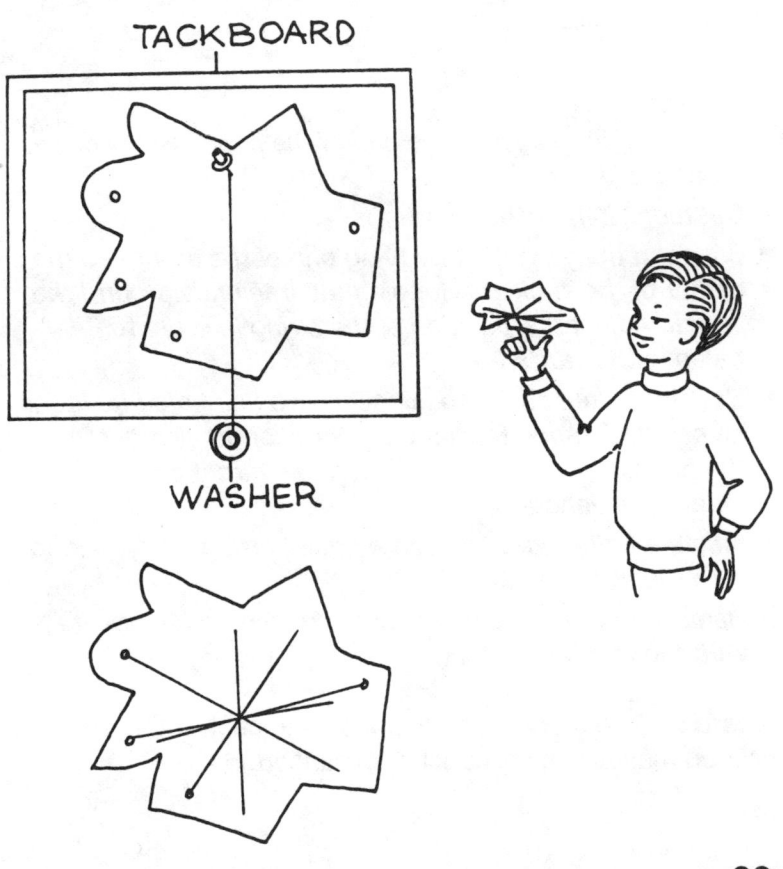

TACKBOARD

WASHER

38. Heavy Air

Purpose To demonstrate that air has weight.

Materials *modeling clay*
pencil
yardstick (meter stick)
3 balloons, 9 in. (23 cm); the balloons must be
the same size
string, 4 ft. (120 cm)
scissors

Procedure

- *Use the clay to secure the end of the pencil to the edge of the table.*
- *Cut four 12-in. (30-cm) strings.*
- *Suspend the yardstick by tying one of the strings to the center of the stick and looping the free ends around the pencil. Adjust the position of the string in order to balance the yardstick.*
- *Use the precut string to suspend two uninflated balloons an equal distance from the center support string. Move the balloons back and forth until the yardstick and the balloons balance.*
- *Inflate a balloon and attach a precut string. Make a loop with the string ends.*
- *Remove one of the uninflated balloons and replace it with the inflated balloon.*

Results The uninflated balloons balance, but the air-filled balloon makes the yardstick unbalanced.

Why? The yardstick balances when the downward pull is the same on both sides of the center support string. Replacing the uninflated balloon with an air-filled balloon puts weight on one side of the stick. The weight of the air inside the balloon increases the downward pull on the stick, causing it to move down on that side.

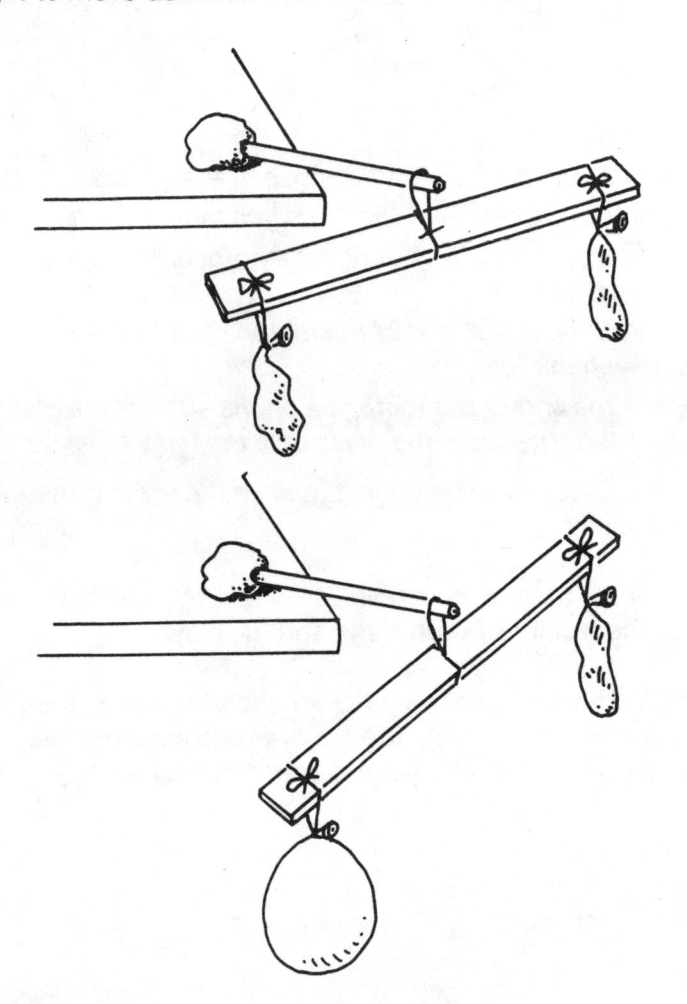

39. Balancing Act

Purpose To determine the center of gravity point.

Materials *2 metal forks*
drinking glass or wide-mouthed jar
modeling clay
1 flat toothpick

Procedure
- *Make a ball of clay about the size of a large marble.*
- *Insert the tip of one of the forks into the clay ball.*
- *Insert the second fork at about a 45-degree angle from the first fork.*
- *Insert the pointed end of the toothpick in the clay between the forks.*
- *Place the end of the toothpick on the edge of the glass. Move it further over the glass until the forks balance.*

   Note: Decrease the angle between the forks if they will not balance.

Results There is one point at which the toothpick supports the weight of both forks and the clay.

Why? The angle of the forks spreads their weight so that there is one place on the toothpick where all of the weight is evenly distributed. This spot is called the *center of gravity*.

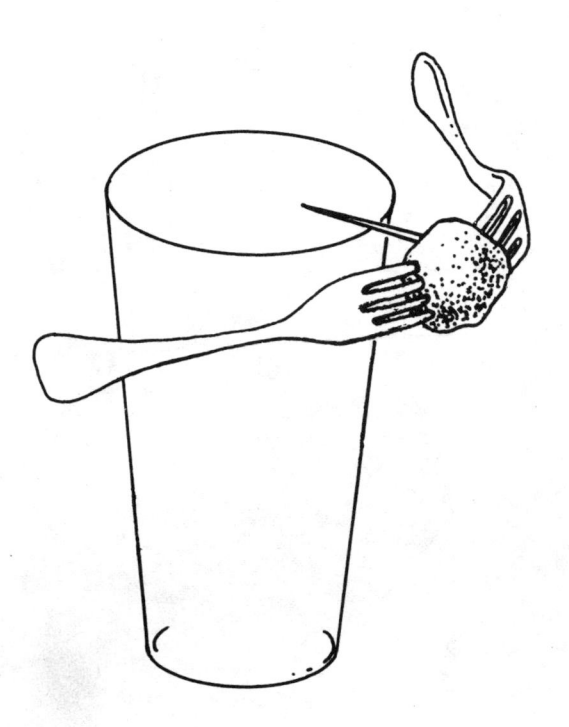

VI
Flight

40. Lift Off

Purpose To demonstrate the effect of a kite's tail.

Materials 1 sheet of notebook paper
scissors
string
ruler
cellophane tape

Procedure

- Cut a 2-in. × 12-in. (5-cm × 30-cm) strip from the sheet of paper.
- Use tape to attach an 18-in. (45-cm) length of string to one end of the strip.
- Hold the free end of the string and whip the paper back and forth in front of you.
- Cut a 1/4-in. × 12-in. (0.5-cm × 30-cm) strip from the paper and attach it with tape to the free end of the wider strip.
- Again move the strip back and forth in front of you.

Results The paper twirls around, but when the small strip is attached the movement is smoother.

Why? The paper moves forward at an angle, causing the air to flow faster over the top side. Fast-moving air has a lower pressure around the moving stream. Thus, more uplift is exerted on the bottom of the strip. The angle of the paper is not constant, causing changes in the pressure along with a turbulent air flow across the strip. These changes make the strip twist and rotate. The paper tail makes the angle more constant. Therefore, there is a smoother flow of air across the paper and less twisting.

97

41. Paper Flop

Purpose To demonstrate the effect of speed on air pressure.

Materials *2 books of equal size*
1 sheet of notebook paper
1 drinking straw
ruler

Procedure

- *Position the books 4 in. (10 cm) apart on a table.*
- *Lay the sheet of paper across the space between the books.*
- *Place the end of the straw just under the edge of the paper.*
- *Blow as hard as you can through the straw.*

Results The paper flops down toward the table when air is blown under it.

Why? Air was pushing equally on all sides of the paper before you blew through the straw. As the speed of a flow of air increases, the sideways pressure of the air decreases. Forcing a stream of fast-moving air under the paper reduces the upward pressure on the paper. The air pushing down on the paper is greater than the air pushing up, thus the paper flops down.

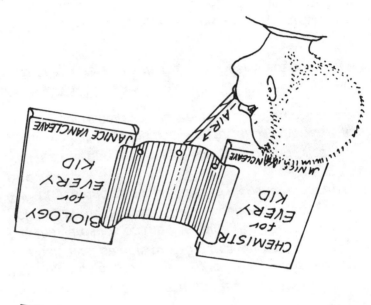

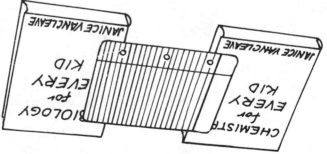

42. Curve Ball

Purpose To demonstrate how a baseball pitcher throws a curve ball.

Materials *golf ball, or any ball of comparable size and weight*
string, 3 ft. (90 cm)
masking tape
yardstick (meter stick)

Procedure

- *Attach the string to the ball with a strip of tape.*
- *Tape the free end of the string to the top edge of a table. The ball must be able to swing freely from the table.*
- *Lay the yardstick on the floor directly under the ball and pointing in the direction that the ball will be swung.*
- *Twist the string about 50 times in a counterclockwise direction.*
- *Pull the ball and the twisted string back and release it.*
- *Allow the ball to swing back and forth. Use the yardstick on the floor to help you observe the direction of the swing.*

Results The ball spins in a clockwise direction as it moves forward. After a few swings, its forward motion curves to the right.

Why? The ball is moving in two directions, around on its own axis and forward. These two directions of motion cause air to flow around the ball in two different directions. As the ball spins clockwise, air is carried with the spinning ball in a clockwise direction. Air pushes in the opposite direction when the ball moves forward. Look at the diagram of the top view of the moving ball. Notice that the arrows

100

are pointing in the same direction on one side of the ball. This indicates that the air is moving in the same direction, which causes an increase in the speed of the air on this side. The air currents are moving in opposite directions on the other side and push against each other, forcing the air to move more slowly on that side. Fast-moving air applies less sideways pressure, so the ball has less pressure on the side with the fast-moving air and more pressure on the side with the slow-moving air. The ball curves to the side with less air pressure. To throw a curve ball, a pitcher must spin the ball as he throws it forward. If he spins it clockwise it will curve to the right, but if he spins it counterclockwise the ball will curve to the left.

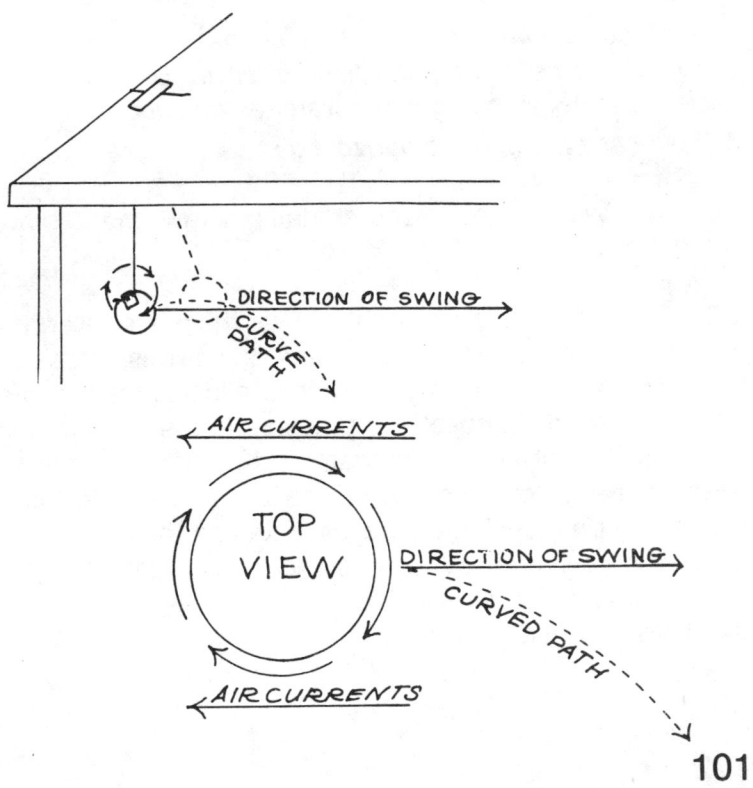

DIRECTION OF SWING

CURVE PATH

AIR CURRENTS

TOP VIEW

DIRECTION OF SWING

CURVED PATH

AIR CURRENTS

43. Swish

Purpose To demonstrate how a perfume sprayer works.

Materials two flexible drinking straws
drinking glass
scissors

Procedure

- Fill the glass with water.
- Cut one straw so that the top of the flexible section stands 1/2 in. (1 cm) above the water's surface. Stand this straw in the water.
- Hold the second straw horizontally so that its end is pointed across the top of the other straw. Use the ridges on the straw standing in the water as a support.
- Blow hard through the horizontal straw.

Results Water rises in the standing straw and is blown out in a mist.

Why? The faster air moves, the lower the pressure around the flow of air. As the air from the straw moves across the top of the standing straw, the pressure inside the standing tube is lowered. Atmospheric pressure in the room pushes down on the surface of the water in the glass, forcing the water to the top of the straw, where it is blown out in a mist. Squeezing on a perfume sprayer produces the same situation. Air is forced across a tube and the perfume rises due to the reduced pressure inside and is thus sprayed outward by the moving air.

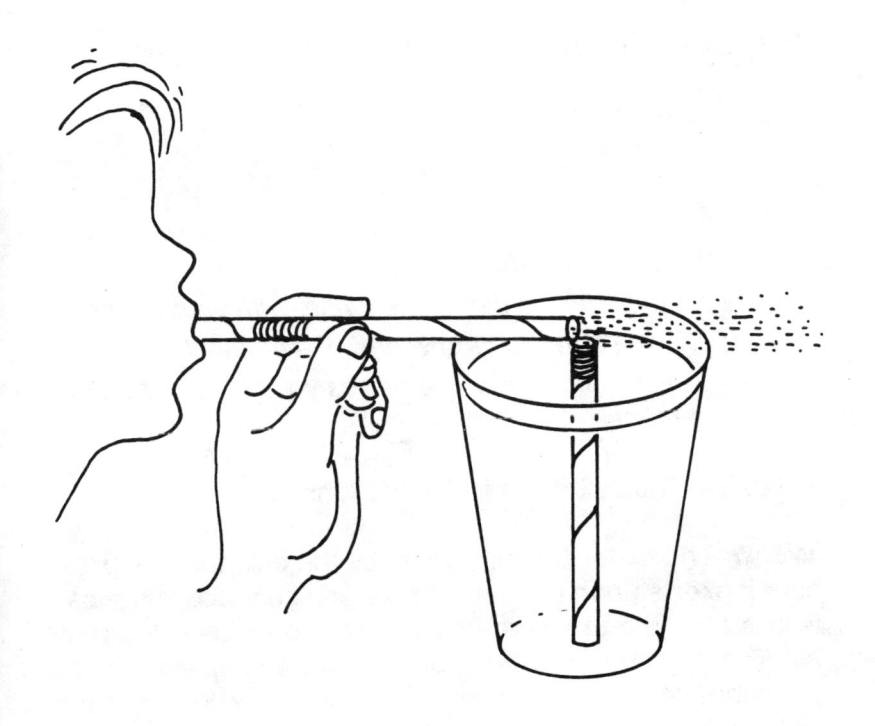

44. Floating Ball

Purpose To demonstrate how air speed affects flight.

Materials *small funnel*
table tennis ball

Procedure
- *Turn the funnel upside down.*
- *Hold the table tennis ball in the funnel with your finger.*
- *Start blowing into the narrow end of the funnel.*
- *Remove your finger from the ball as you continue to blow into the funnel.*

Results The ball floats inside the funnel.

Why? The faster the air passes by the ball, the less pressure it exerts upon the ball. The air pressure above the ball is less than the pressure under it, so the ball is held up by air. The pressure of moving air explains the upward lift on the wings of aircraft. When the air flows faster over the top of the wing than below, there is an upward push called *lift*.

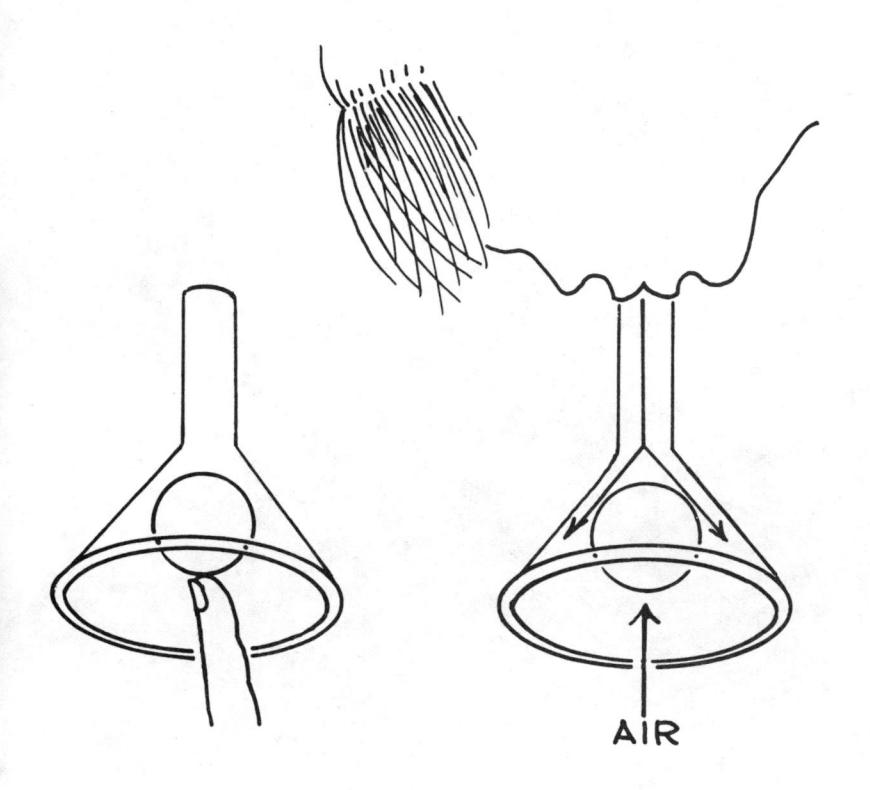

AIR

VII

Simple Machines

45. Ramp

Purpose To demonstrate that a winding mountain road is an inclined plane.

Materials *sheet of paper*
pencil
scissors
ruler
cellophane tape

Procedure
- *Cut a 5-in. (13-cm) square from the paper.*
- *Draw a diagonal line across the square and cut across the line.*
- *Color the longest edge of the paper triangle with the pencil.*
- *Tape the triangle to the pencil as shown in the diagram.*
- *Wind the paper onto the pencil.*

Results The colored side of the triangle is shaped like a ramp. Wrapping the paper around the pencil makes it look like a winding road or a screw.

Why? An *inclined plane* is a tilted flat surface. A winding road gradually gets higher. If the road could be unwound, it would look like the paper triangle with the road running up the long, colored side. It is true that it would take longer to travel up the winding road than it would to walk up the mountain's side, but it takes much less effort to travel the longer distance. The inclined plane is a simple machine that makes a job easier. In this case, the job is climbing a mountain.

108

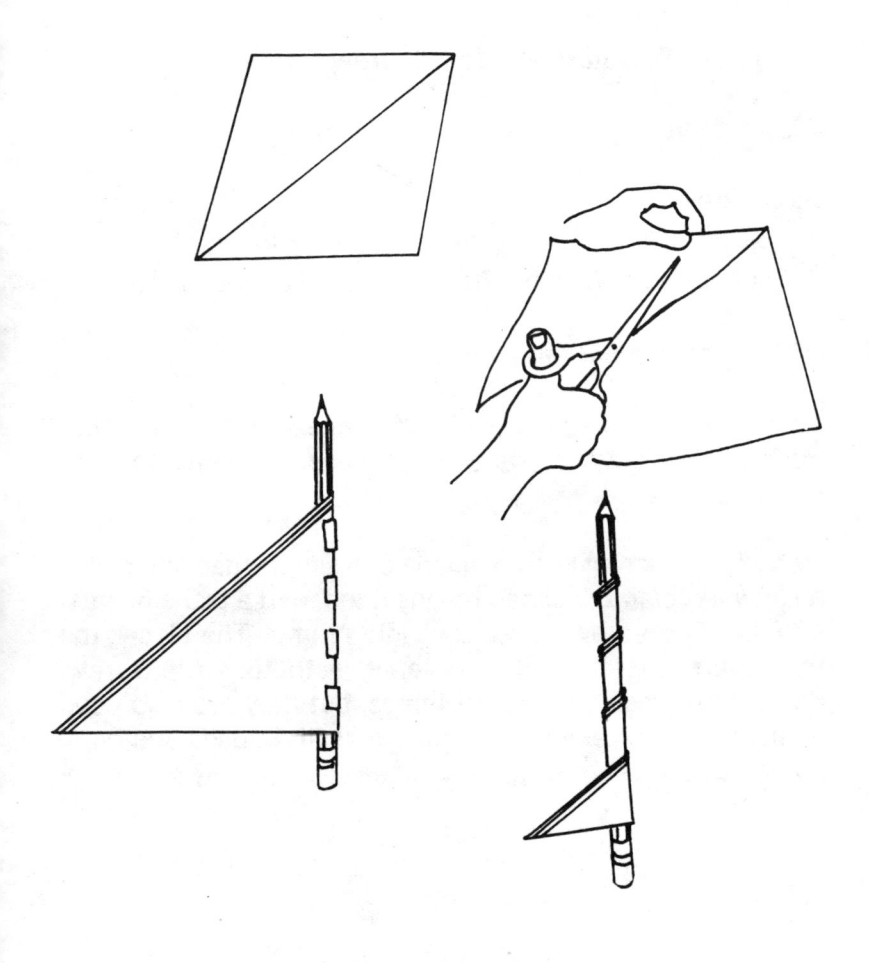

46. Lifter

Purpose To determine how a screw works.

Materials *large screw*

Procedure
- *Hold the head of the screw with one hand.*
- *Put two fingernails on the first ridge at the tip of the screw.*
- *Turn the head of the screw.*

Results As the screw turns, the ridges spiral downward. Your fingers stay in place and this unique winding road moves past them.

Why? The screw is an example of a simple machine called an *inclined plane.* Inclined means tilted, and a plane is a flat surface. The screw is like a winding ramp. The closer the treads are together, the easier it is to turn the screw. Screws are used to connect things, but they are also used to lift things; for example, screw jacks lift houses and cars. Machines are used to make a job easier.

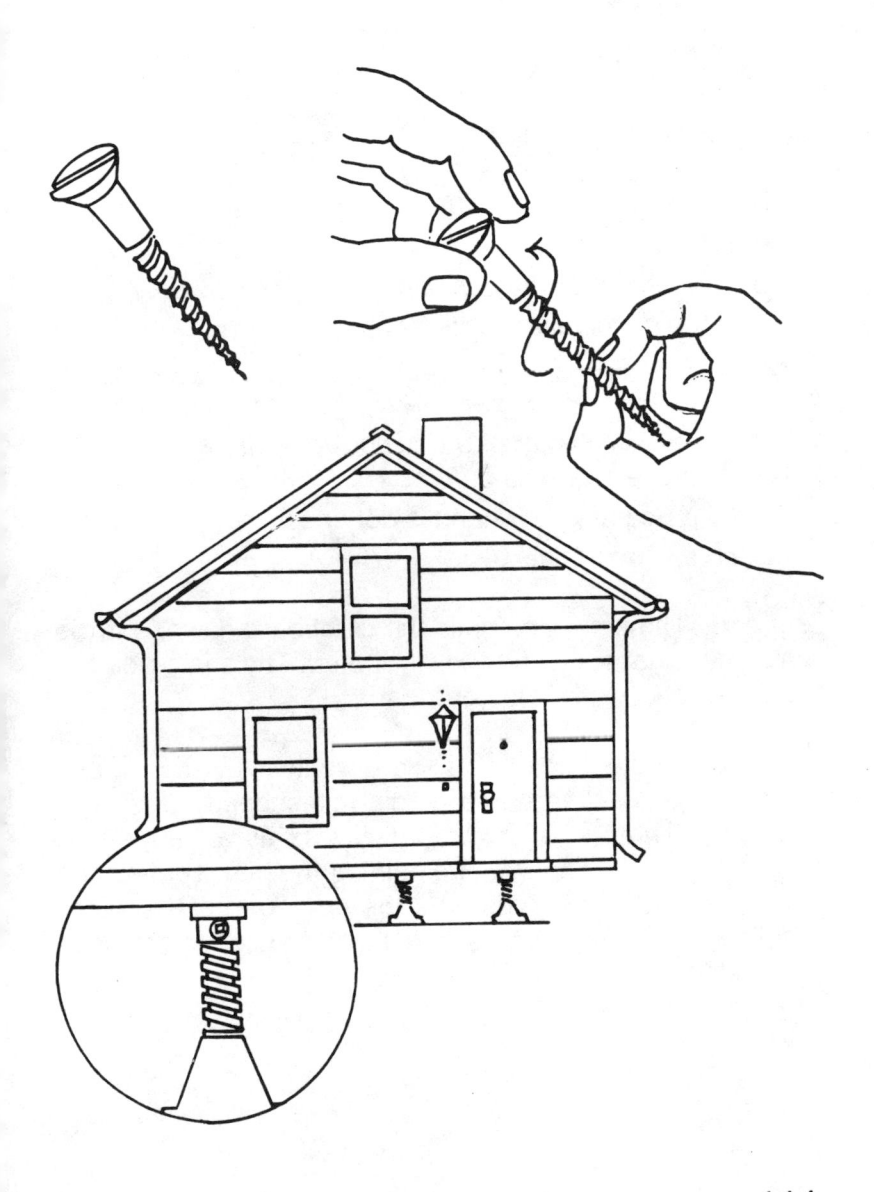

111

47. Wedge

Purpose To demonstrate how a wedge works.

Materials *pencil*
piece of cardboard

Procedure
- *Hold the eraser of the pencil on top of the cardboard.*
- *Press the pencil down against the cardboard. Press hard.*
- *Observe what happens.*
- *Press the sharp point of the pencil against the cardboard, as hard as before.*
  Caution: Be sure your hand is not under the cardboard.
- *Observe what happens.*

Results The eraser presses the cardboard down, but the eraser does not stick into the cardboard. The sharp point of the pencil sticks into the cardboard.

Why? The pencil point acts like a wedge. A wedge is any material that is tapered to a thin, pie-shaped edge. The tapered, pointed end of the pencil acts as a wedge and sticks into the cardboard. The point of the pencil enters the paper first and makes a path for the larger part of the pencil to follow. An example of a simple machine used to split logs is a wedge.

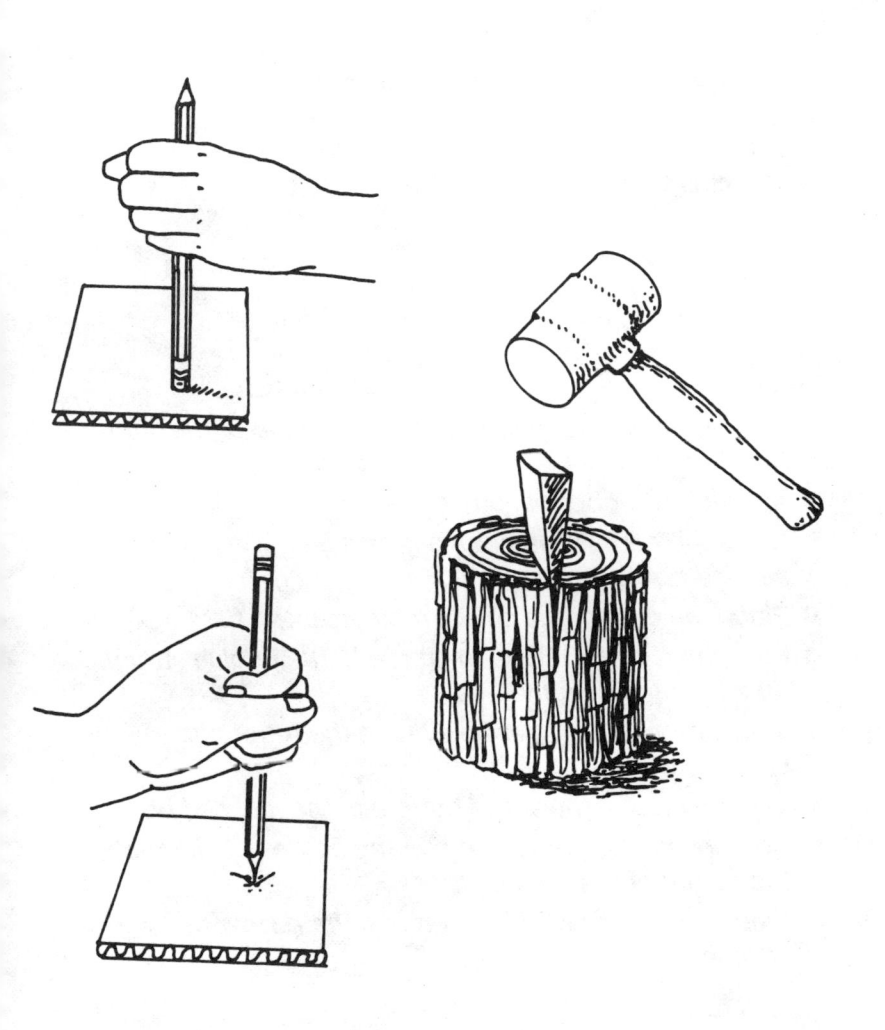

113

48. Inclined

Purpose To determine if an inclined plane makes work easier.

Materials *string, 12 in. (30 cm)*
rubber band
2 books
glue bottle, 8 oz. (236 ml)
ruler

Procedure

■ *Tie the center of the string around the top of the glue bottle.*

■ *Attach the rubber band to the string.*

■ *Raise the end of one book and rest it on top of the second book.*

■ *Place the glue bottle on the table near the flat book.*

■ *Hold the rubber band and lift the bottle of glue straight up and place it on the book.*

■ *Measure with the ruler how much the rubber band stretches.*

■ *Lay the glue bottle at the bottom of the inclined book.*

■ *Hold the rubber band and pull the bottle until it reaches the height of the flat book.*

■ *Measure with the ruler how much the rubber band stretches.*

Results The rubber band stretches the most when the bottle is lifted straight up.

Why? You lifted the glue bottle to the same height each time, but lifting it straight up took more effort because you

114

were holding the entire weight of the bottle. The books form a simple machine called an *inclined plane* that helps to support some of the weight of the bottle. You had to pull the bottle up the incline a longer distance to reach the height of the book, but it took much less effort, as shown by the rubber band that stretched less.

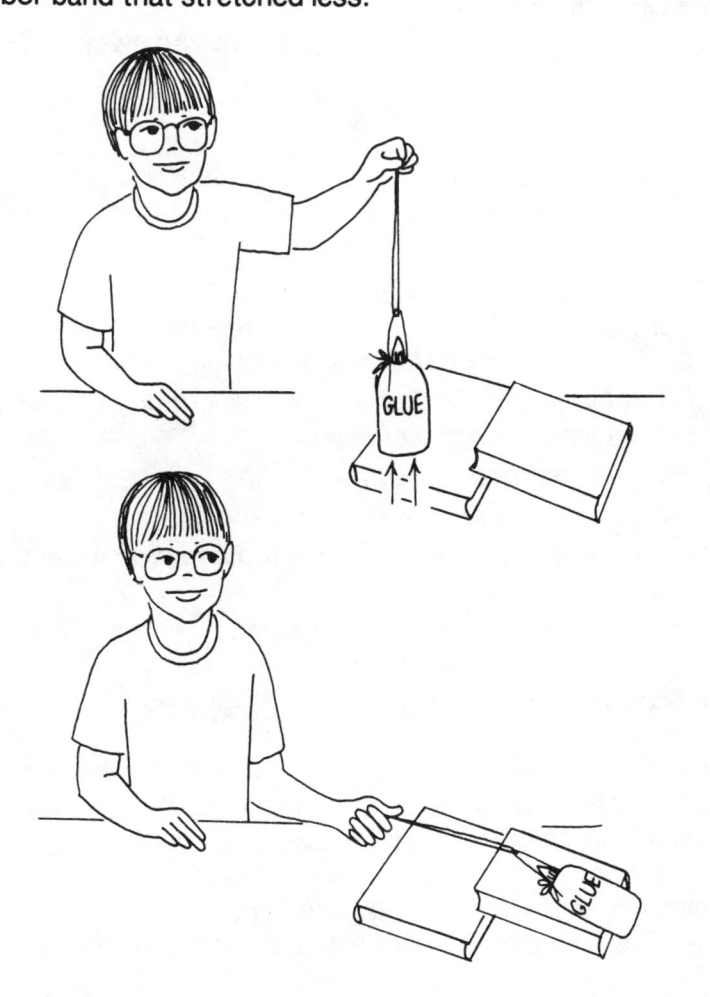

49. Pumps

Purpose To demonstrate how a wheel can be used as a pump.

Materials *eye dropper*
glass or wide-mouthed jar with a 4-in. (10-cm) opening
poster board
round toothpick
scissors
helper

Procedure

■ *Cut a circle with a 4-in. (10-cm) diameter from the poster board, using the glass to outline the circle.*

■ *Insert about one-third of the toothpick through the center of the circle to form a top.*

■ *Fill the eye dropper with water.*

■ *Ask your helper to spin the paper top.*

■ *Hold the eye dropper above the top and drop water onto the spinning disk.*

■ *Observe the movement of the water.*

Results The water sprays out in all directions.

Why? The falling water is thrown away from the spinning disk. If you could slow down the motion of the water, you would see that the water leaves the spinning top in a straight line. There is an outward pull on all spinning materials; if free to move, they are thrown off in a straight line. This knowledge was used to design a simple machine to pump water. Water flowing onto a spinning metal disk creates a wheel pump that very effectively pumps water from one spot to another.

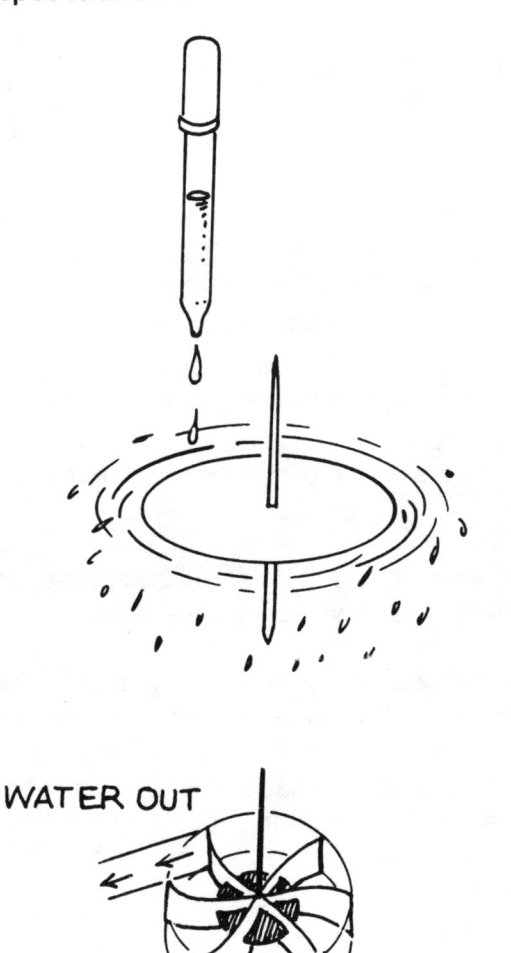

WATER OUT

WATER IN

50. Levers

Purpose To demonstrate the effectiveness of a *lever*.

Materials *4 books*
2 pencils

Procedure
- *Stack the books.*
- *Put your little finger under the edge of the bottom book in the stack and try to lift the books.*
- *Place one pencil under the edge of the bottom book in the stack.*
- *Place the second pencil under the first pencil near the book.*
- *Push down on the end of the second pencil and try to lift the books.*

Results It is very difficult to lift the books with your finger alone, but easy when two pencils are used.

Why? The pencils form a *lever*. One of the pencils acts as a *fulcrum* (a point of rotation), and the second pencil is the lever arm. As the distance from where you push down to the fulcrum increases, the easier it is to lift the object on the opposite end. Levers are simple machines that multiply the force that you apply. This makes moving or lifting large objects easier.

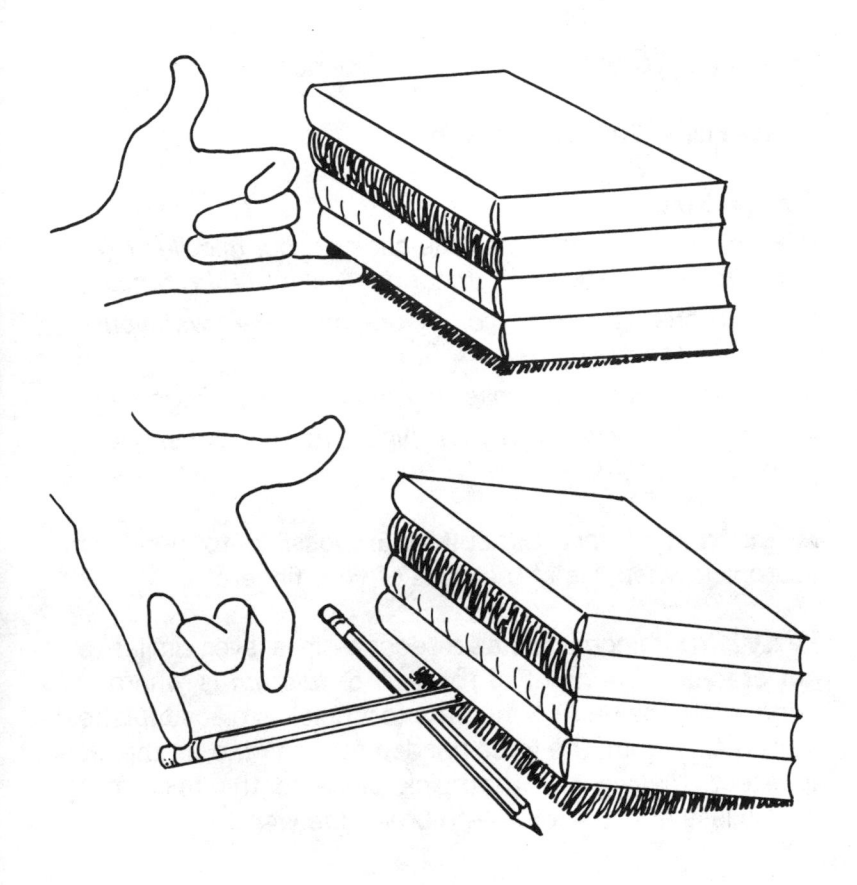

51. Weakling

Purpose To demonstrate a second-class lever.

Materials *2 round toothpicks*

Procedure
- *Place a toothpick across the back of your middle finger at the first knuckle and under the first and third finger.*
- *Try to break the toothpick by pressing down with your first and third fingers.*
- *Move the toothpick closer to the tips of your fingers.*
- *Again push down with your fingers to try to break the toothpick.*

Results It is very difficult or impossible to break the toothpick when it is at the ends of your fingers.

Why? Your fingers act as a second-class lever similar to a nut cracker. The point of rotation or *fulcrum* is where the fingers join the hand. When the toothpick is placed furthest from the fulcrum, the force needed to break the toothpick is greatest. Placing the toothpick close to the fulcrum requires less effort—force—to break the wood.

EFFORT
FORCE

52. Best Spot

Purpose To determine if the position of the lever's fulcrum affects the lever.

Materials *ruler*
pencil
30 pennies

Procedure
- *Make a lever by laying the ruler across the pencil.*
- *Move the pencil under the 4-in. (10-cm) mark on the ruler.*
- *Place 10 pennies between the end of the ruler and the 1-in. (2.5-cm) mark.*
- *Add and record the number of pennies needed on the opposite end of the ruler to lift the 10 pennies.*
- *Move the pencil to the 8-in. (20-cm) mark.*
- *Again position 10 pennies on the end of the ruler.*
- *Add and record the number of pennies needed on the opposite end of the ruler to lift the 10 pennies.*

Results It took many more pennies to lift the 10 pennies when the pencil was at the 8-in. (20-cm) mark.

Why? The ruler and pencil form a simple machine called a *lever*. The pencil acts as the *fulcrum* that the lever rotates around. The effort is the amount of force needed to make the object move. In this experiment, the effort was the stack of pennies needed to lift the 10 pennies. The longer the distance from the point of effort to the fulcrum, the less effort needed to move the object on the other end. You

122

experience this on a see-saw when the heavy person moves to the center in order to balance the lighter person sitting on the other end. The lever can make lifting or moving an object an easy task if the fulcrum is placed close to the object to be moved.

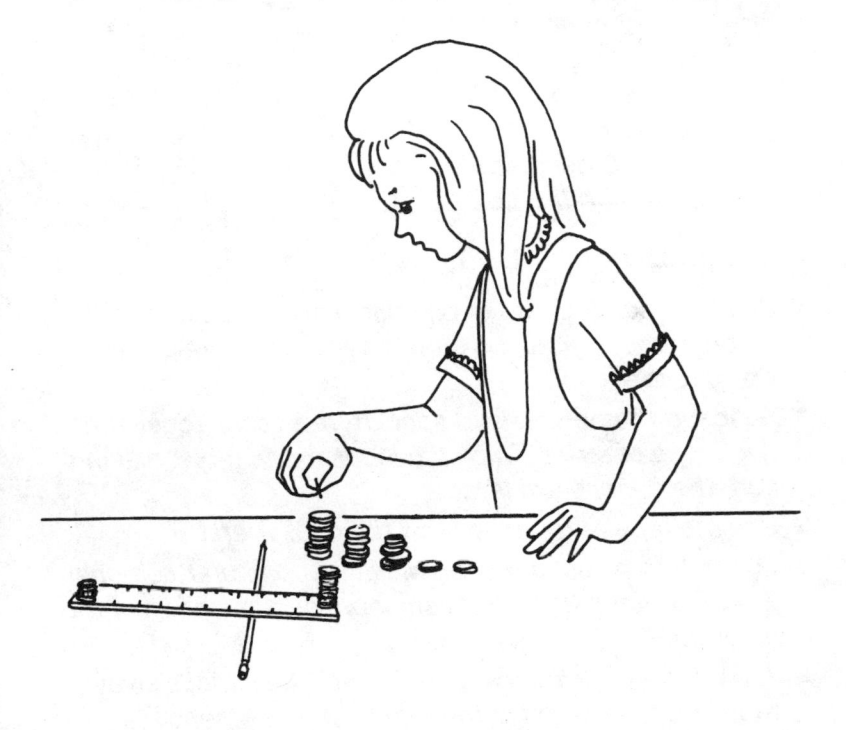

53. Wheel and Axle

Purpose To demonstrate how machines make work easier.

Materials *empty thread spool*
2 pencils
string
scissors
2 paper cups
20 pennies
marking pen

Procedure

- *Stick the pointed end of one pencil into each end of the empty thread spool. Be sure the pencils fit snugly and do not slide.*
- *Suspend the pencils and spool from a table edge with two loops of string. Tape the loops to the table, making sure the pencils are level.*
- *Use the tape and pen to label the cups A and B.*
- *Punch two holes at the top of each cup and attach one 24-in. (60-cm) string to each paper cup as shown in the diagram.*
- *Tape cup A's string to a pencil. Turn the pencils away from you to wind all of the thread onto the pencil.*
- *Tape cup B's thread to the outside of the thread spool.*
- *Turn the pencils toward you to wind cup B's thread onto the spool.*
- *Place 10 pennies in cup A.*
- *Cup B should be at its top position. Add pennies to cup B one at a time until it starts to slowly move downward.*
- *Observe the distance that both cups move.*

124

Results It takes 6 to 7 pennies in cup B to raise the 10 pennies in cup A. Cup B moves a greater distance than cup A.

Why? The distance around the spool is farther than the distance around the pencils. This makes the cup attached to the spool move farther every time the spool and the pencils revolve. Machines make work easier. This spool and pencil machine, called a *wheel and axle*, needs the force of only 6 to 7 pennies to lift the load of 10 pennies.

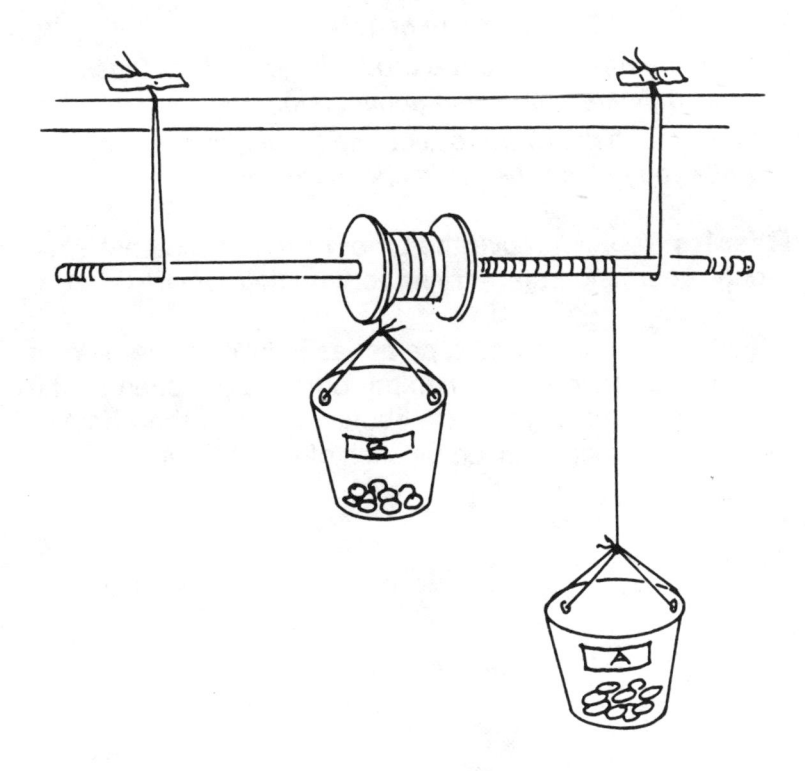

54. Tug of War

Purpose To demonstrate how easily things are moved with a machine.

Materials *two brooms*
rope or strong cord, 9 ft. (3 m)
2 helpers

Procedure
- *Tie the rope to one broom handle*
- *Wrap the rope around the broom handles three times while they are being held about 20 in. (50 cm) apart.*
- *Have your helpers try to keep the broom handles apart while you pull on the loose end of the rope.*

Results You can move the broom handles together even though your helpers are trying to keep them apart.

Why? The brooms and rope act as a pulley system. Your force is multiplied by the number of ropes attached to the brooms; therefore, you have about five times the effort or force that is being exerted by each of your helpers.

VIII

Inertia

55. Crash!

Purpose To demonstrate that moving objects have inertia.

Materials *piece of modeling clay, size of a walnut*
2 rulers
small toy car that can roll on the ruler
masking tape
pencil
2 books about 1 in. (2.5 cm) thick

Procedure

- *Raise one end of the ruler and place it on one of the books.*
- *Tape the other end of the ruler to a table.*
- *Tape the pencil perpendicular to and about two car lengths from the end of the ruler.*
- *Make a clay figure similar to a snowman.*
- *Flatten the bottom of the clay figure and gently sit it on the hood of the toy car. You want the clay figure to fall off the car easily, so do not press the clay against the car.*
- *Position the car with its clay figure at the top of the raised ruler.*
- *Release the car and allow it to roll down the ruler and collide into the pencil.*
- *Use the second ruler to measure how far the clay figure falls from the car.*
- *Repeat the procedure several times before raising the ruler by adding the second book.*
- *Keep a record of how far the clay figure falls from the car.*

Results The car with the clay figure moves down the ruler. The car stops when it hits the ruler, but the clay figure continues to move forward. Raising the ruler causes the clay figure to fall farther from the car.

130

Why? As the car rolls down the ruler, its speed increases. The clay figure has the same speed as the car. When the car hits the pencil, the force of impact stops the car, but the clay figure is free to continue moving forward until some force stops it. Raising the height of the ruler causes the car to reach a higher speed before it hits the pencil, and so the clay figure also moves at a higher speed. The faster the clay figure moves, the farther it flies before the force of the air molecules brings it to a stop.

The car and clay figure both have *inertia,* a resistance to a change in motion. Once started, both continue to move until some outside force acts against them, causing them to stop. The pencil stopped the car's motion, and the air molecules slowed the clay figure until its forward motion stopped and gravity pulled the clay figure down.

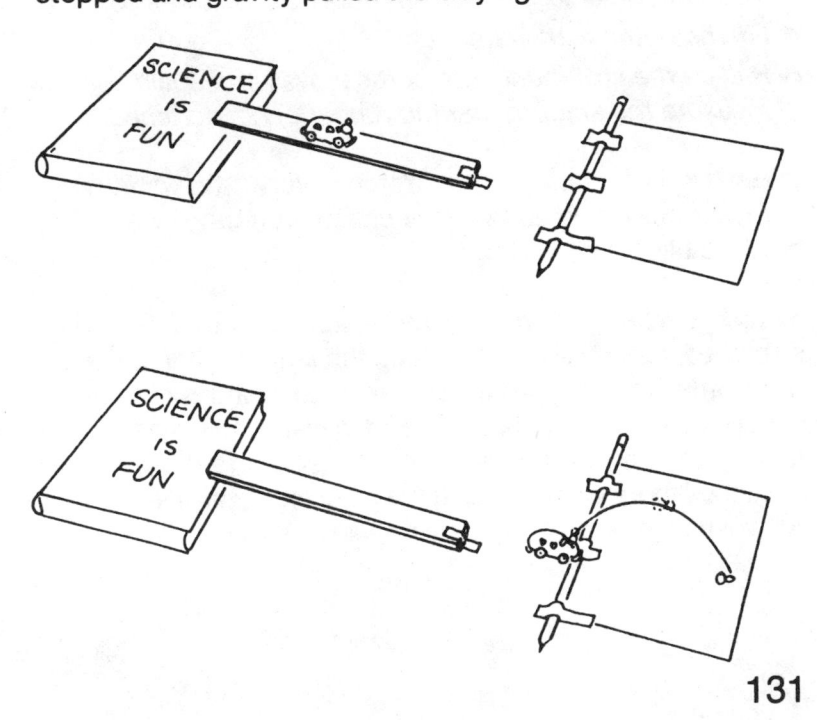

56. More

Purpose To demonstrate the effect of weight on inertia.

Materials *plastic soft-drink bottle, 2 qt. (2 liter)*
rubber band
string, 12 in. (30 cm)
scissors
ruler

Procedure
- *Tie the string to the rubber band.*
- *Place the rubber band around the bottom of the bottle.*
- *Pull on the string until the bottle starts to move.*
- *Measure the amount that the rubber band stretches.*
- *Fill the bottle with water.*
- *Pull on the string until the bottle moves and again measure the amount that the rubber band stretches.*

Results The rubber band stretches very little when moving the empty bottle, but stretches much more when the bottle is filled with water.

Why? Inertia is the resistance to motion. The rubber band stretched very little when moving the empty bottle because the bottle has very little resistance to being moved. The water-filled bottle is heavier and resists movement more than does the empty bottle. As a result, the rubber band stretched more when moving the heavier, water-filled bottle. As weight increases, *inertia* increases.

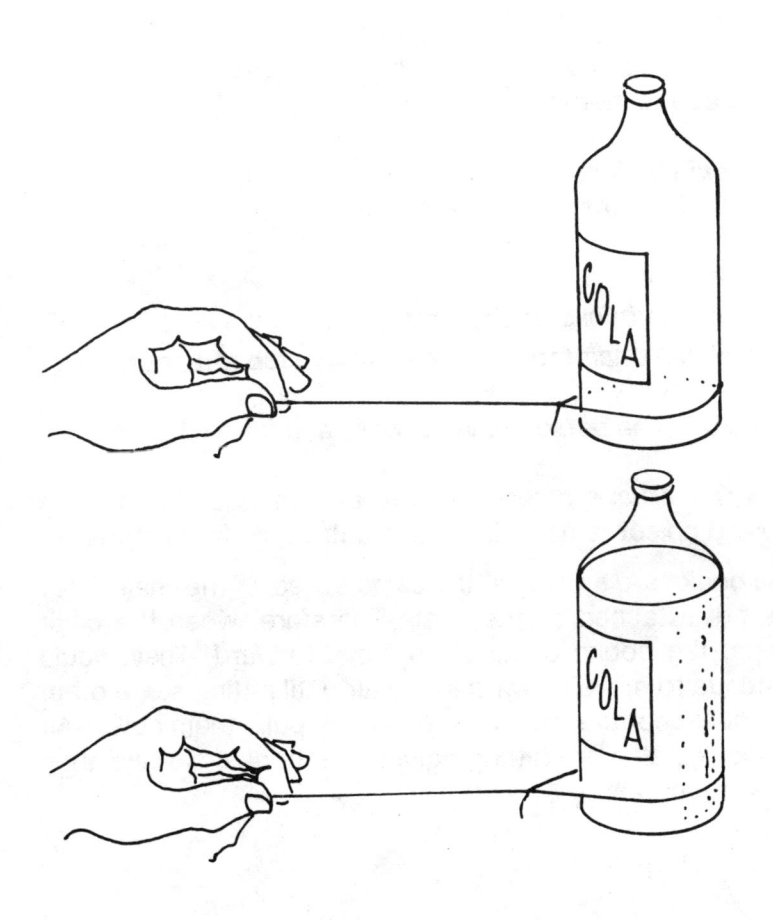

57. Plop!

Purpose To demonstrate that an object continues to move due to inertia.

Materials *5 books*
chair or cart with rollers

Procedure
- *Stack the books on the edge of the chair's seat.*
- *Push the chair forward then quickly stop the chair.*

Results The books move forward and fall to the floor.

Why? *Inertia* is a resistance to any change in motion. A moving object remains in motion until some force stops it.

The books are moving at the same speed as the chair. They are not attached to the chair. Therefore, when the chair stops, the books continue to move forward. They would continue to move forward in the air until hitting some other object except that the force of *gravity* pulls them down. Air molecules also are hitting against the books, slowing their forward motion.

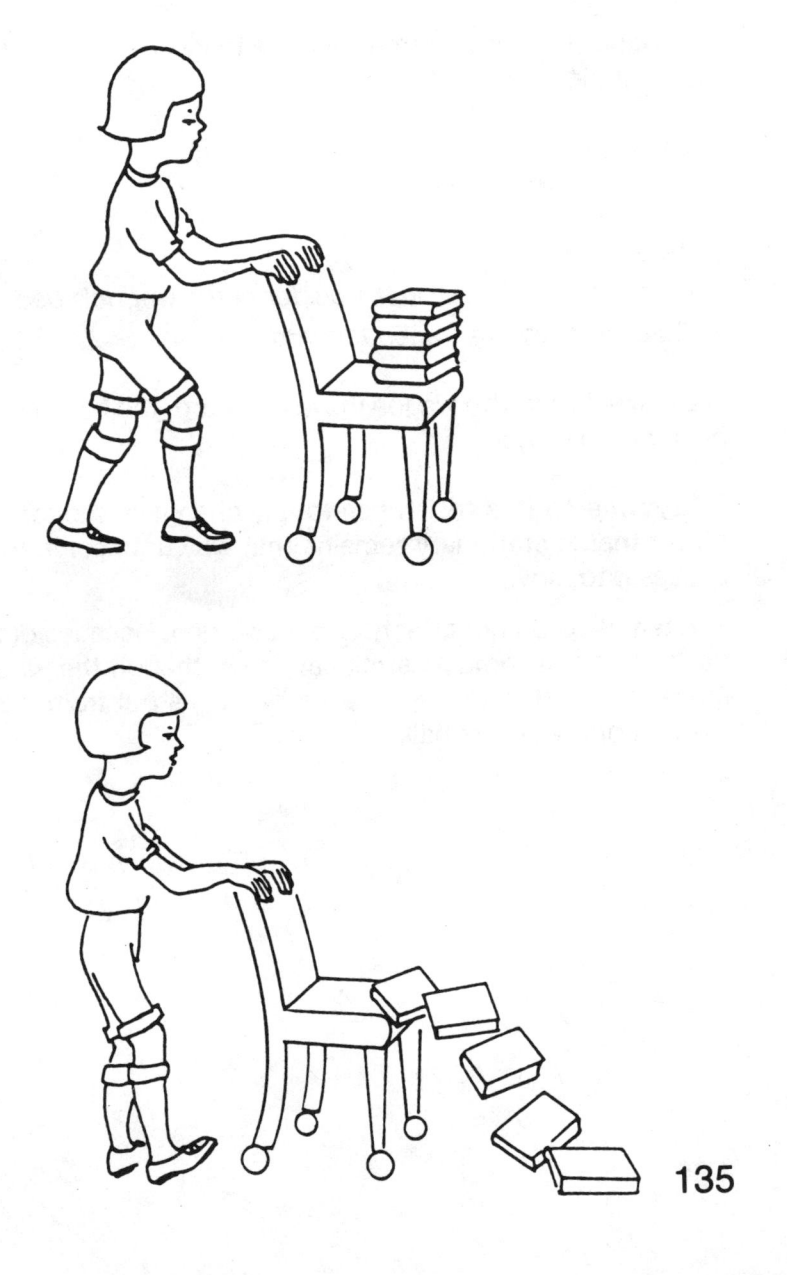

58. Oops!

Purpose To demonstrate that due to inertia an object remains stationary.

Materials *wagon*
tennis ball

Procedure

- *Place the tennis ball in the center of the wagon's bed.*
- *Quickly move the wagon forward.*

Results When the wagon moves forward, the ball hits the back of the wagon.

Why? *Inertia* is a resistance to any change in motion. An object that is stationary remains that way until some force causes it to move.

The tennis ball is not attached to the wagon. Because of the ball's inertia, it remains stationary even though the wagon moves forward. The wagon actually moves out from under the stationary tennis ball.

59. Thump!

Purpose To demonstrate how forces affect inertia.

Materials *drinking glass*
index card
clothespin

Procedure
- *Place the index card over the mouth of the glass.*
- *Place the clothespin on top of the card so that it is centered over the glass.*
- *Quickly and forcefully thump the card straight forward with your finger.*
- *Repeat the experiment several times.*

Results The clothespin falls straight into the glass about half of the time; the other half of the time it flips over, landing upside down in the glass.

Why? Your finger applies force to the card, moving it forward. The card moves so quickly that it translates very little force to the clothespin. The pin falls straight down due to the pull of gravity when the card no longer supports it. If you do not hit the card straight forward with enough force, it pulls the bottom of the pin forward and gravity pulls the top of the pin down, causing the pin to flip before it lands.

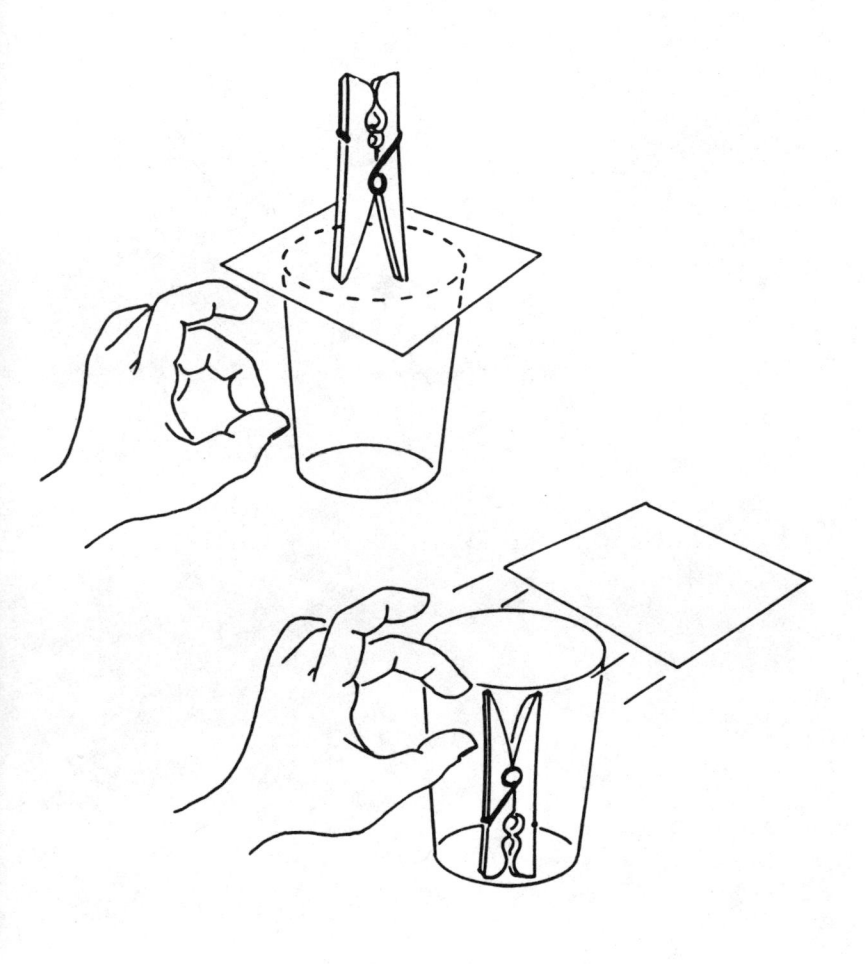

IX
Motion

60. Balloon Rocket

Purpose To demonstrate how unbalanced forces produce motion.

Materials yardstick (meter stick)
drinking straw
scissors
string
balloon, 9 in. (23 cm)
2 chairs
masking tape

Procedure

- Cut a 4-in. (10-cm) piece from the drinking straw.
- Cut about 3½ ft. (4.5 m) of string.
- Thread the end of the string through the straw piece.
- Position the chairs about 4 ft. (4 m) apart.
- Tie the string to the backs of the chairs. Make the string as tight as possible.
- Inflate the balloon and twist the open end.
- Move the straw to one end of the string.
- Tape the inflated balloon to the straw.
- Release the balloon.

Results The straw with the attached balloon jets across the string. The movement stops at the end of the string or when the balloon totally deflates.

Why? Newton's Law of Action and Reaction states that when an object is pushed, it pushes back. When the balloon

142

was opened, the walls of the balloon pushed the air out. When the balloon pushed against the air, the air pushed back and the balloon moved forward, dragging the straw with it. The string and straw keep the balloon rocket on a straight course.

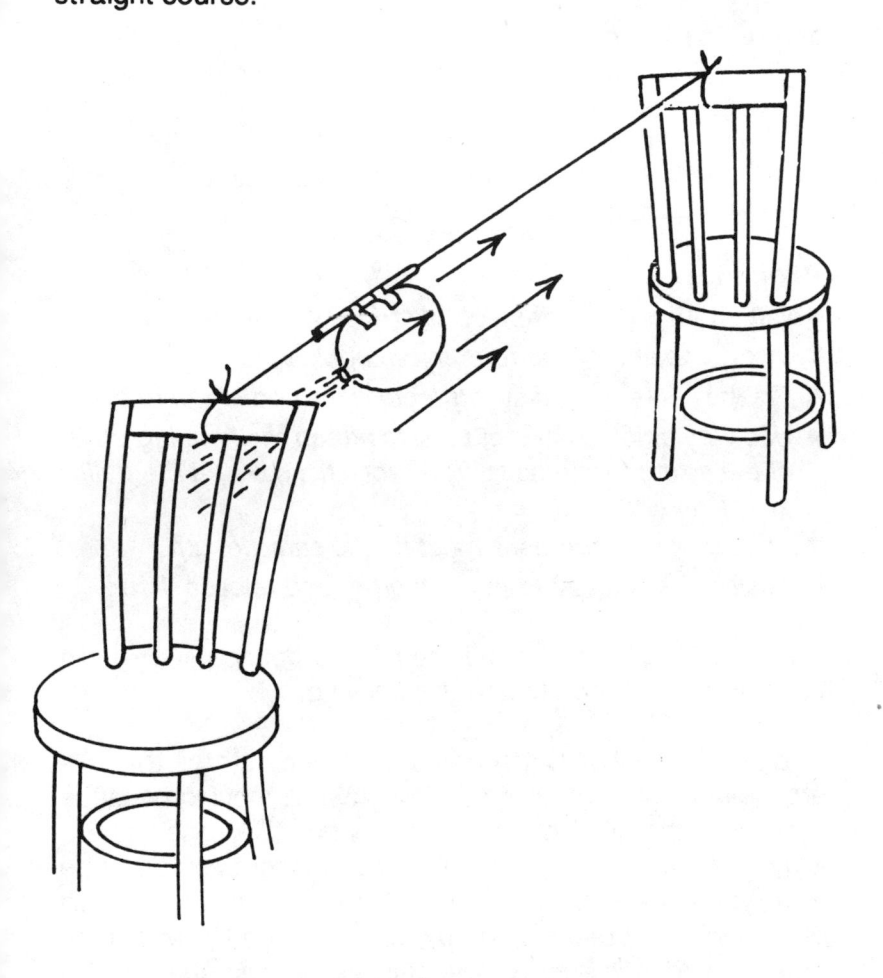

143

61. Bonk!

Purpose To determine what happens to energy after it is used.

Materials *ruler*
string, 24 in. (60 cm)
duct tape
book
2 small rubber balls of equal size
ruler

Procedure

- *Cut a 24-in. (60-cm) piece of string.*
- *Insert one end of the ruler into a book.*
- *Tie the center of the string around the end of the ruler.*
- *Use very small pieces of tape to attach the hanging ends of the string to the balls. The tape sticks best if the balls are clean and oil-free.*
- *The strings on the balls must be the same length.*
- *Pull the balls away from each other and release them.*

Results The balls continue to hit and bounce away from each other until they finally stop moving.

Why? The *Law of Conservation of Energy* states that energy is never lost or created. Things start moving because they have energy and stop moving when they lose the energy. The energy to raise the balls to a higher position came from you. Releasing the balls allowed the energy due to their height, or *potential energy,* to be changed to energy of motion, or *kinetic energy.* The balls exchanged energy with each other when they collided. Energy has a magnitude and

144

a direction. Upon collision, the balls received a specific amount of energy. They also received a push in the opposite direction, resulting in backward motion after contact. The motion energy changes into heat energy as the balls hit each other and rub against air molecules. The balls stop moving when their energy is gone, but the air around the balls is warmer because it received the heat energy. Energy is never lost, only changed to another form or transferred to another object.

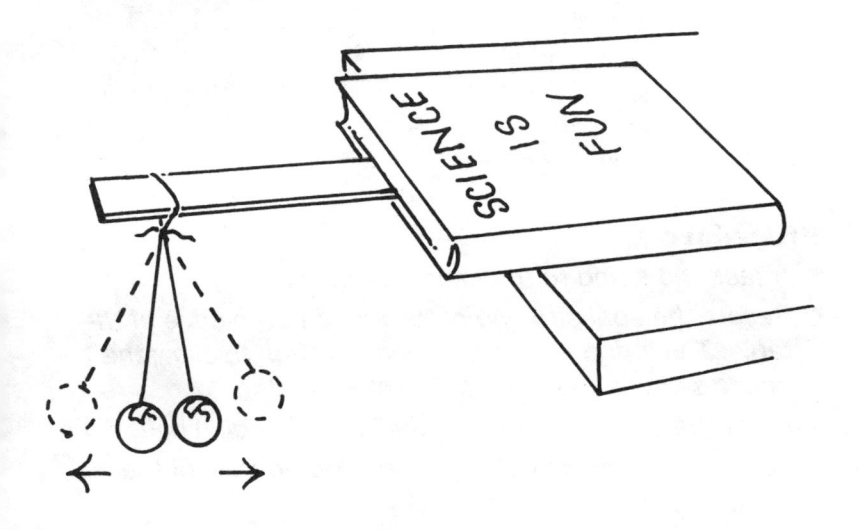

62. Loser

Purpose To determine the effect that mass has on kinetic energy.

Materials *table*
small pail
masking tape
paper
pencil
wooden block
string, 1 yd. (1 m)
scissors
clay

Procedure

■ *Attach the string to the handle of the pail.*
■ *Secure the opposite end of the string to the edge of the table. The string needs to be long enough to allow the pail to swing about 1 in. (2.5 cm) above the floor.*
■ *Place the paper on the floor under the hanging pail.*
■ *Position the wooden block on the floor in front of the hanging pail.*
■ *Pull the pail back and allow it to swing into the block. Mark the position that the block has moved to on the paper.*
■ *Again position the wooden block on the floor in front of the hanging pail.*
■ *Place large pieces of clay in the pail to increase its weight.*
■ *Pull the pail back to the same position as before, then allow the pail to swing into the wooden block.*
■ *Mark the position of the block on the paper.*

146

Results The wooden block moves farther when struck by the heavier pail.

Why? The pail was raised to the same height each time, keeping the speed constant. The mass of the pail increased when the clay was added. Kinetic energy (energy of motion) is increased as the mass of a moving object is increased. The heavier pail had more energy when it struck the wooden block, and thus it pushed the block farther along the paper.

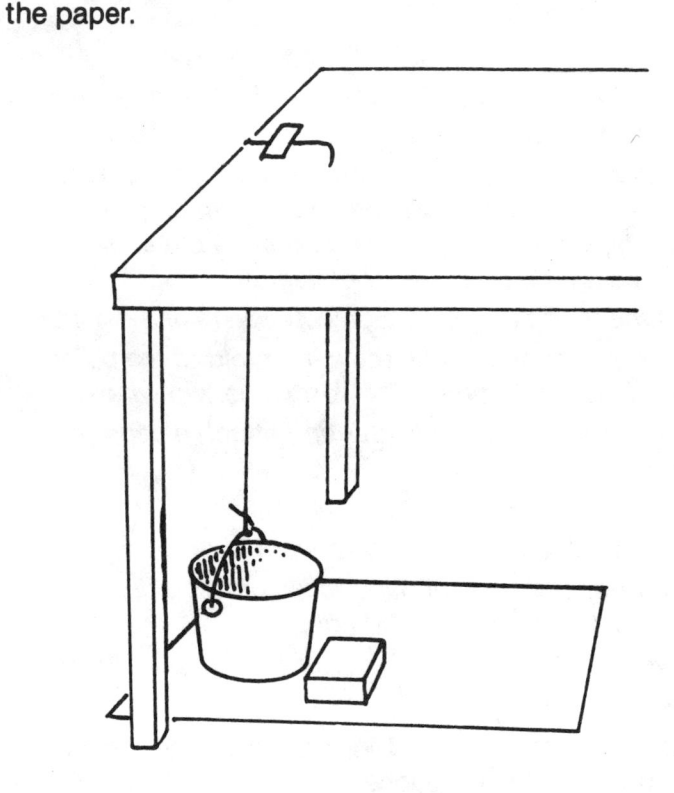

63. Helicopter

Purpose To determine how weight affects the rotation speed of a paper helicopter.

Materials 3 paper clips
pencil
notebook paper
scissors
ruler

Procedure

- Fold and cut one sheet of paper in half lengthwise.
- Take one of the halves and fold it in half lengthwise.
- Use a ruler to draw a triangle on one edge of the paper. The base of the triangle will be 1 in. (3 cm) long and one side will be between the 4-in. and 6-in. (9 cm and 14 cm) marks on the ruler. See the diagram.
- Cut out the triangle. Cut through both layers of the paper.
- Open the paper and cut up the center fold to the point indicated on the diagram. This forms the two wings.
- Fold the tabs toward the center and attach a paper clip to the bottom.
- Fold the wings in opposite directions.
- Hold the helicopter above your head and drop it.
- Add different numbers of paper clips one at a time and drop the plane after each addition.
- Observe the speed of rotation after each paper clip is added.

Results The rotation speed increases as the weight increases, but a point is reached where additional weight pulls down with such force that the wings move upward and the plane falls like any falling object.

148

Why? As the paper falls, air rushes out from under the wings in all directions. The air hits against the body of the craft, causing it to rotate. Increasing the weight by adding paper clips causes the helicopter to fall faster, and the amount of air hitting the craft's body increases. This increase in air movement under the wings increases the rotation speed.

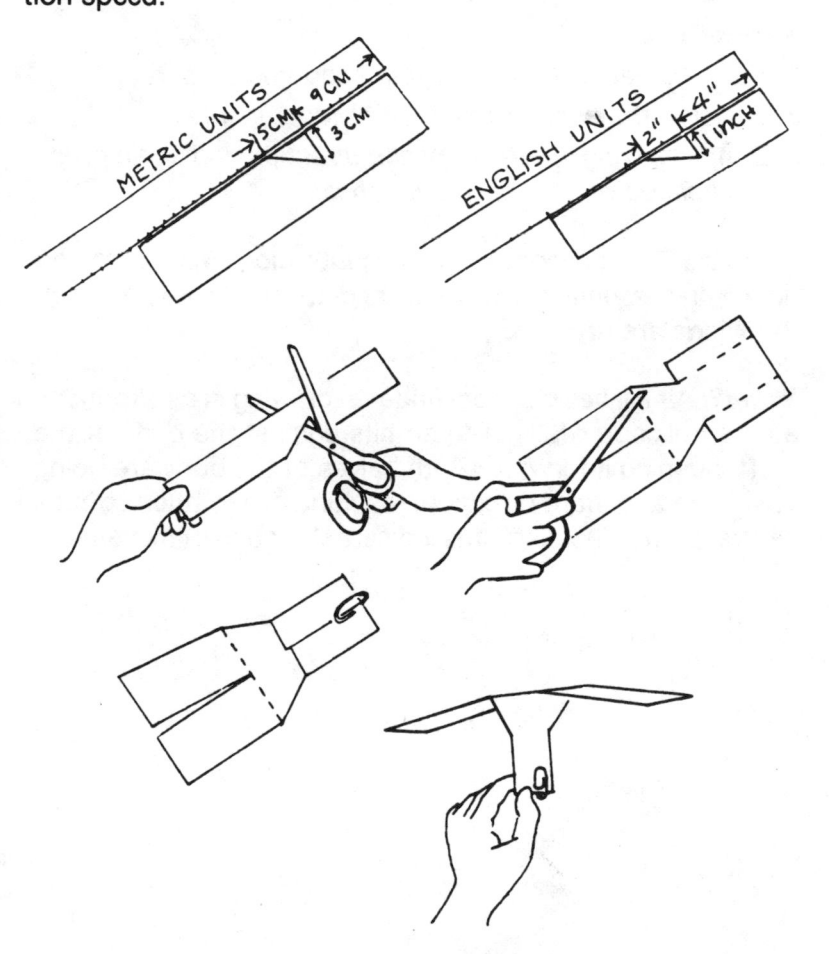

64. Right or Left?

Purpose To determine how wing position affects the direction of a paper helicopter's rotation.

Materials *paper helicopter from Experiment 63*

Procedure
- *Hold the helicopter above your head and drop it.*
- *Observe the direction that the helicopter spins.*
- *Bend the wings in the opposite direction and again drop the helicopter from above your head.*

Results The helicopter spins counterclockwise when the right wing is bent toward you and turns clockwise when the wings are reversed.

Why? Air rushes out from under each wing in all directions as the helicopter falls. The air hits against the body of the craft, pushing it forward. Both halves of the body are being pushed in a forward direction, resulting in a rotation about a central point. The diagrams indicate direction of movement.

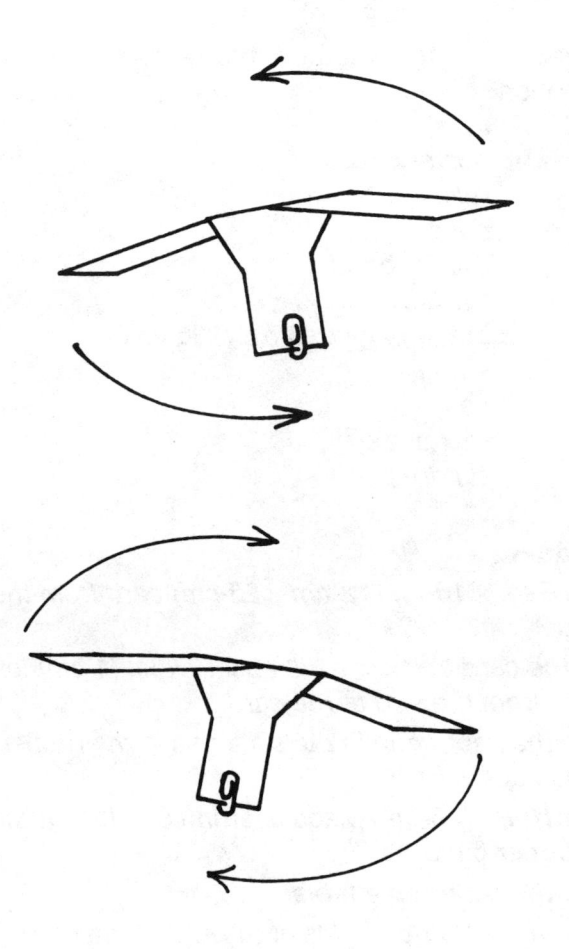

65. How Far?

Purpose To determine how the texture of a surface affects motion.

Materials poster board
ruler
paper clip
rubber band
scissors
bottle of glue, 8 oz. (236 ml)
string
pen
wax paper
sandpaper

Procedure

- Cut a 5-in. × 10-in. (12-cm × 25-cm) card from the poster board.
- Fold the card and cut a slit about 1/2 in. (1 cm) long 2 in. (5 cm) from the end of the card.
- Place the paper clip in the slit and slip the rubber band on the paper clip.
- Cut a 10-in. (25-cm) piece of string and loop it through the rubber band.
- Place the paper on a table.
- Position the bottle of glue at the end of the card.
- Gently pull on the string to straighten the rubber band.
- Mark the card at the end of the rubber band and label this mark START.
- Pull on the string until the card begins to move.
- Note how much the rubber band stretches.

- *Tape a sheet of wax paper and a sheet of sandpaper to the table.*
- *Move the card with the glue bottle across the wax paper and sandpaper by pulling on the string.*
- *Observe how much the rubber band stretches each time.*

Results The rubber band stretched the least when the card was placed on the wax paper and stretched the most on the sandpaper.

Why? The weight of the glue bottle pushes the card down against the surface it sits on. The card is more easily pulled across the wax paper than across the sandpaper because of *friction.* Friction is a force that pushes against a moving object, causing the object to stop moving. Frictional force increases with the roughness of the surfaces moving against each other. The surface of the wax paper is smoother than that of the sandpaper or table, and thus applies less frictional force to the card.

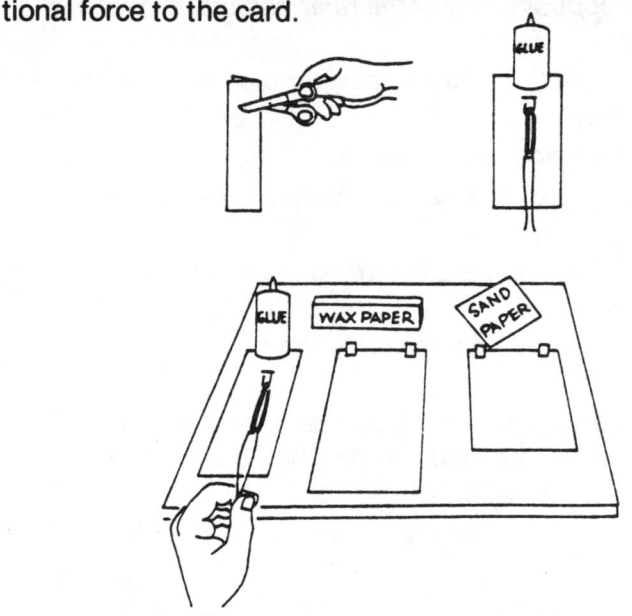

66. Energy Change

Purpose To demonstrate the effect that height has on the energy of a moving object.

Materials book
pencil
ruler with a center groove
paper cup, 8 oz. (236 ml)
scissors
marble

Procedure

- *Cut a 1-in. (2.5-cm) square section from the top of the paper cup.*
- *Place the cup over the ruler. The end of the ruler should touch the back side of the cup.*
- *Raise the opposite end of the ruler and rest it on the pencil.*
- *Place the marble in the center groove of the ruler at the ruler's highest end.*
- *Release the marble and observe the cup.*
- *Raise the end of the ruler and rest it on the edge of the book.*
- *Again position the marble in the groove at the ruler's highest end.*
- *Release the marble and observe the cup.*

Results The cup moves when the marble strikes it. The cup moved an additional distance when the ruler rested on the book.

Why? Objects at rest have *potential energy.* The higher the object sits above the ground, the greater is its potential energy. When objects fall or roll down an incline, their potential energy changes into *kinetic energy* —energy of motion. Increasing the height from which the marble rolled gave it more energy, causing it to strike the cup with more force.

67. Roller

Purpose To determine how surface area affects friction.

Materials *string, 24 in. (60 cm)*
rubber band
2 large books
10 round marking pens
ruler

Procedure

- *Stack the books on a table.*
- *Tie a string around the bottom book.*
- *Attach the string to the rubber band.*
- *Move the stack of books by pulling on the rubber band.*
- *Measure how far the rubber band stretches.*
- *Place the 10 marking pens under the stack of books.*
- *Move the books by pulling on the rubber band.*
- *Observe how far the rubber band stretches.*

Results The rubber band stretches more when the bottom book sits flat against the table than when it is placed on the pens.

Why? *Friction* is a force that tries to stop motion. Friction increases as the amount of surface area in contact with the surface increases. The pens reduce friction between the table and the bottom book by decreasing the amount of surface that the book touches.

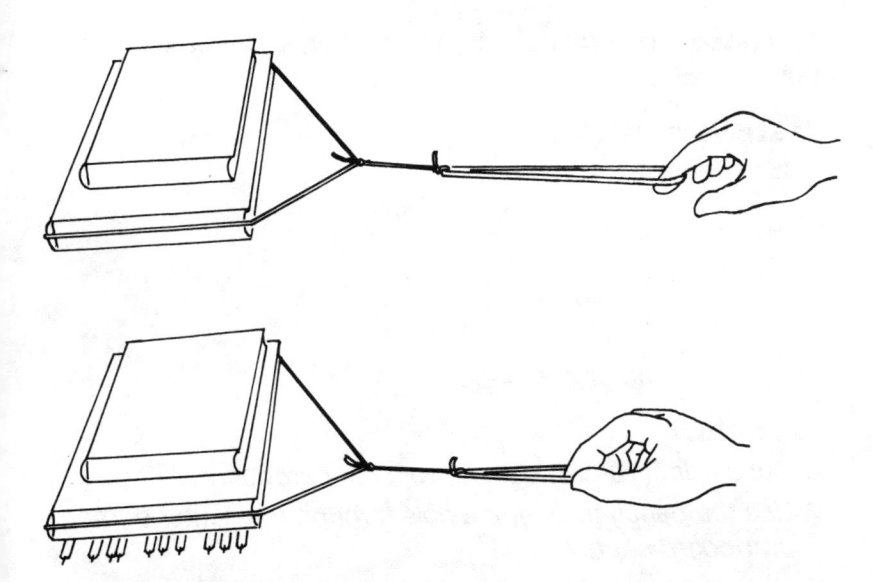

68. Air Car

Purpose To demonstrate how motion is affected by friction.

Materials cardboard
pencil
empty thread spool
balloon, 9 in. (23 cm)
ruler
scissors
glue
notebook paper

Procedure

- Cut a 4-in. (10-cm) square from the cardboard.
- Use the pencil to punch a hole through the center of the cardboard square.
- Glue the empty thread spool over the hole in the cardboard. Be sure the hole in the spool lines up with the hole in the cardboard.
- Place a bead of glue around the base of the spool.
- Cut and glue a circle of paper over the top end of the thread spool. Allow the glue to dry for several hours.
- Use a pencil to punch a hole in the paper circle to line up with the hole in the spool.
- Place the cardboard on a smooth surface such as a table.
- Push the cardboard and observe its motion.
- Inflate the balloon and twist the end.
- Stretch the opening of the balloon over the top of the spool.
- Untwist the balloon and give the cardboard a little push and observe its motion.

158

Results The air car moves very little without the balloon, but easily skims across the table when the air flows through the spool.

Why? The air flowing from the balloon through the holes forms a thin layer of air between the cardboard and table. This air layer reduces friction (a force that tries to stop movement), allowing the car to quickly move across the table.

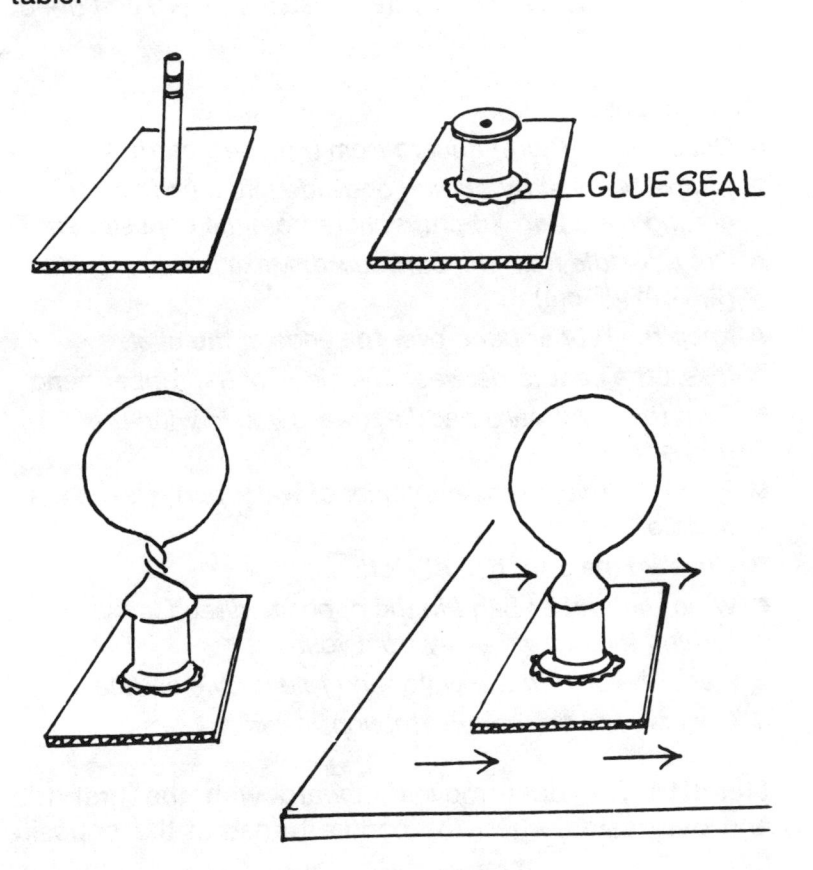

GLUE SEAL

69. Paddle Boat

Purpose To demonstrate Newton's Law of Action and Reaction.

Materials cardboard
rubber band
scissors
container of water at least 4 in. (10 cm) deep
ruler

Procedure

- Cut a 4-in. (10-cm) square from the cardboard.
- Shape the boat by cutting one side into a point and cutting out a 2-in. (5-cm) square from the opposite end.
- Cut a paddle from the cardboard. Make it 1 in. × 2 in. (2.5 cm × 5 cm).
- Loop the rubber band over the ends of the boat.
- Insert the paddle between the sides of the rubber band.
- Turn the cardboard paddle toward you to wind the rubber band.
- Place the boat in the container of water and release the paddle.
- Observe the direction of motion.
- Wind the rubber band in the opposite direction by turning the paddle away from you.
- Place the boat in the water and release the paddle.
- Observe the direction of motion.

Results The boat moves forward with the first trial and backwards when the paddle turned in the opposite direction.

160

Why? *Newton's Law of Action and Reaction* states that when an object is pushed, it pushes back with an equal and opposite force. Winding the paddle caused it to turn and hit against the water. When the paddle pushed against the water, the water pushed back and the boat moved. The boat moved in the opposite direction to the paddle, changing direction when the paddle direction changed.

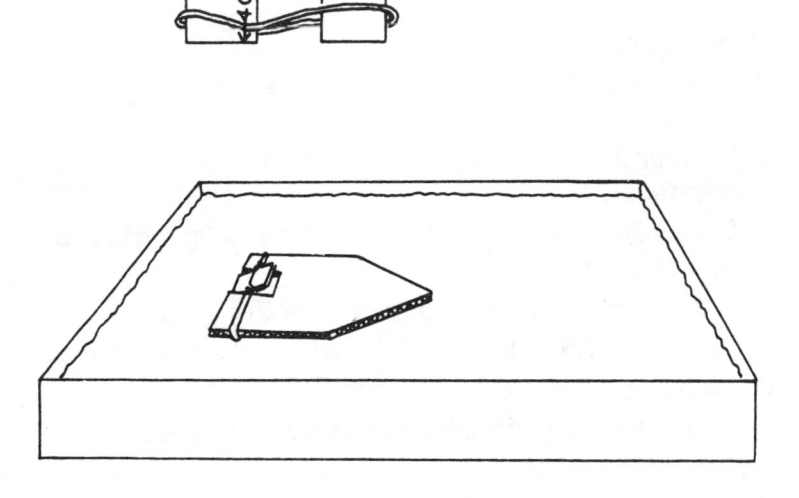

70. Wind Brake

Purpose To demonstrate the effect of friction on motion.

Materials string
2 paper clips
empty styrofoam thread spool
ruler
cellophane tape
stiff paper (index card will work)
scissors
knitting needle

Procedure

- Ask an adult to cut four slits at right angles in the thread spool.
- Cut four 3 in. × 1 ½ in. (7.5 cm × 4 cm) cards from the stiff paper.
- Cut a 16-in. (40-cm) length of string and tape it to the side of the spool.
- Attach two paper clips to the end of the string.
- Put the knitting needle through the center of the spool.
- Wind the string around the spool.
- Hold the knitting needle and observe the speed of the unwinding string.
- Insert the four paper cards in the slits in the spool.
- Wind the string around the spool.
- Hold the knitting needle and observe the speed of the unwinding string.

Results The spool turns more slowly when the paper cards are in place.

162

Why? *Gravity* causes the paper clips to fall and pull the attached string with them. As the string unwinds, it spins the spool. *Friction* is the resistance to motion. There is friction between the paper cards and the molecules of gas in the air. Air pushes against the paper cards as the spool spins, reducing the speed of the turning spool.

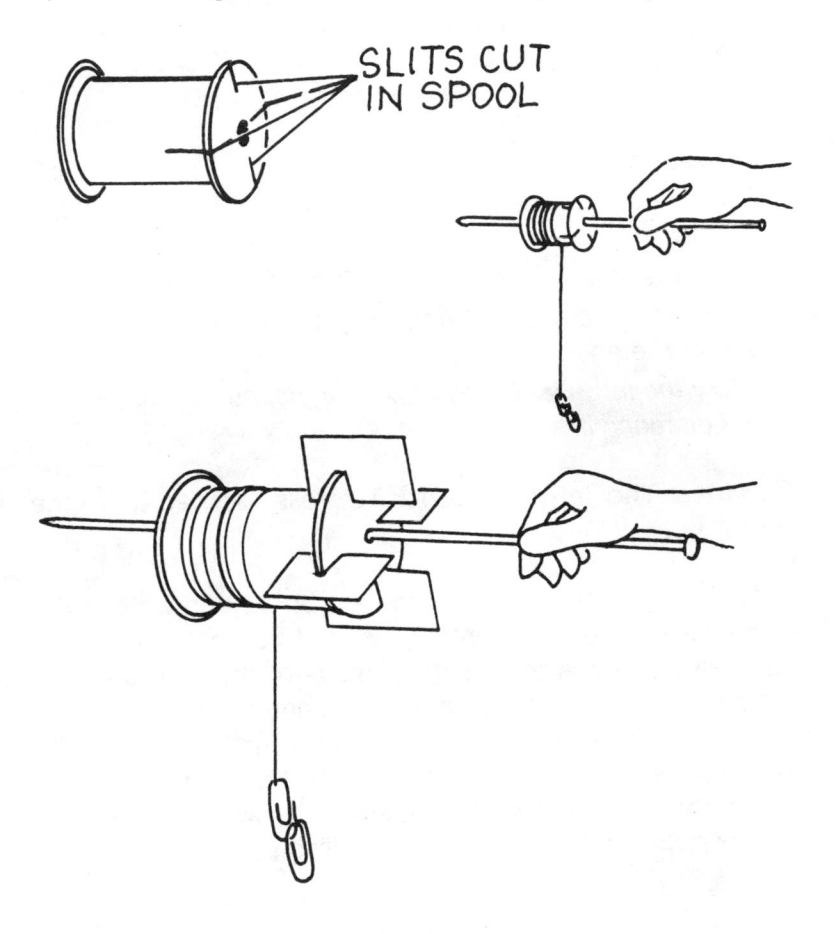

SLITS CUT IN SPOOL

71. Farther

Purpose To determine the effect of weight on the energy of a moving object.

Materials *2 books*
1 small round jar
1 large round jar
yardstick (meter stick)

Procedure

- *Set the edge of one book on the second book to form an incline.*
- *Place the small jar at the top of the incline.*
- *Measure the distance from where the jar stops to the end of the book.*
- *Allow the large jar to roll down the incline and measure the distance it rolls.*

Results The larger, heavier jar rolls farther than the lighter jar.

Why? Since the friction of the air, book, and floor is the same for both experiments, it will not be considered. The major difference is the weight of the two jars. As the weight of the rolling object increases, its energy increases. The energy of the separate rolling jars is calculated by multiplying each weight by the height of the incline. Keeping the height the same—constant—makes the weight of each jar responsible for the change in the rolling distance.

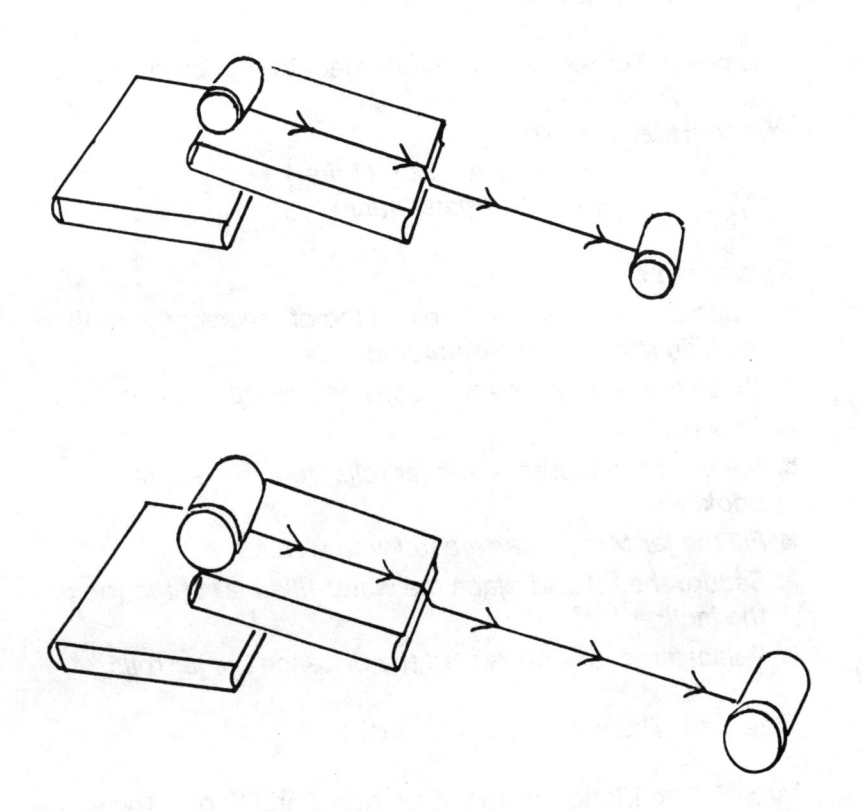

72. Wobbler

Purpose To demonstrate the effect of friction on motion.

Materials *2 books*
1 jar with lid, 1 qt. (1 liter)
yardstick (meter stick)

Procedure

- *Place the edge of one book on top of the second book to form an incline. (See diagram.)*
- *Close the empty jar and place it at the top of the incline.*
- *Release the jar.*
- *Measure the distance the jar rolls from the end of the book.*
- *Fill the jar three-quarters full with water.*
- *Secure the lid and place the water-filled jar at the top of the incline.*
- *Release the jar and record the distance the jar rolls.*

Results The empty jar rolls farther.

Why? The friction from the air, book, and floor is the same during both experiments; thus, they do not have to be considered. In Experiment 71, it was discovered that heavier objects roll farther. The water makes the jar heavier, but the water hits against the inside of the jar as it rolls down the incline, increasing the friction inside the jar. It takes more energy to move this jar with the wobbling water inside.

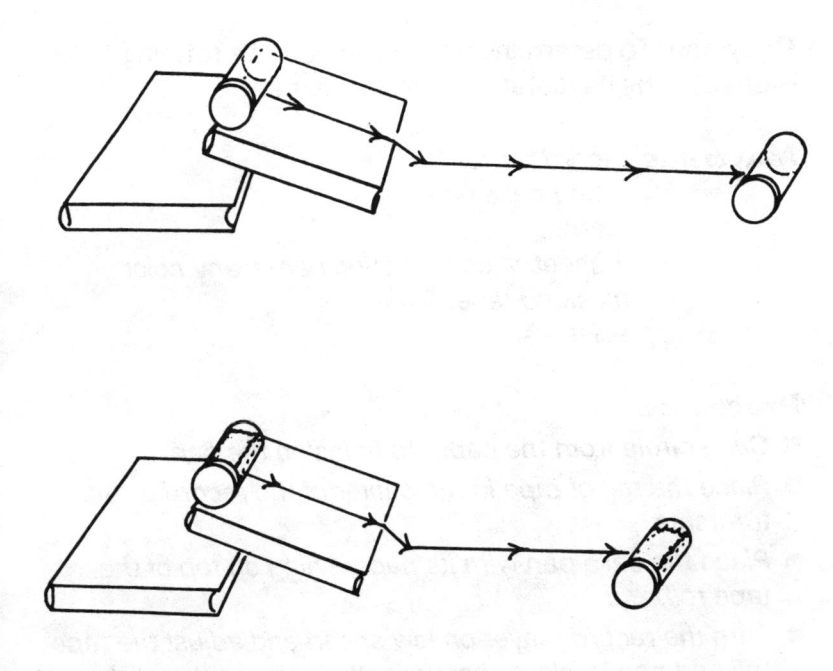

73. Spinner

Purpose To determine how the speed of a rotating object is affected by its distance from the center.

Materials *record player*
round cake pan
marble
1 sheet of construction paper, any color
masking tape, 1 roll
scissors

Procedure

- *Cut a circle from the paper to fit inside the pan.*
- *Place the roll of tape in the center of the record player turntable.*
- *Place the cake pan with its paper lining on top of the tape roll.*
- *Turn the record player on low speed and adjust the tape roll and pan to place them exactly in the center of the turntable.*
- *Increase the speed to high.*
- *Place the marble in different places in the spinning pan.*
- *Observe the speed at which the marble moves.*
- *Place the marble exactly in the center of the rotating pan.*

Results The marble moves quickly when placed near the side of the pan. The marble stays in place when placed in the center of the pan.

Why? The rotating motion of the pan moves the marble outward. The speed of the marble depends on how close it is to the rotating center. The closer to the center, the more slowly the marble moves outward, until a point is reached where the marble's outward speed equals zero; that point is the exact center.

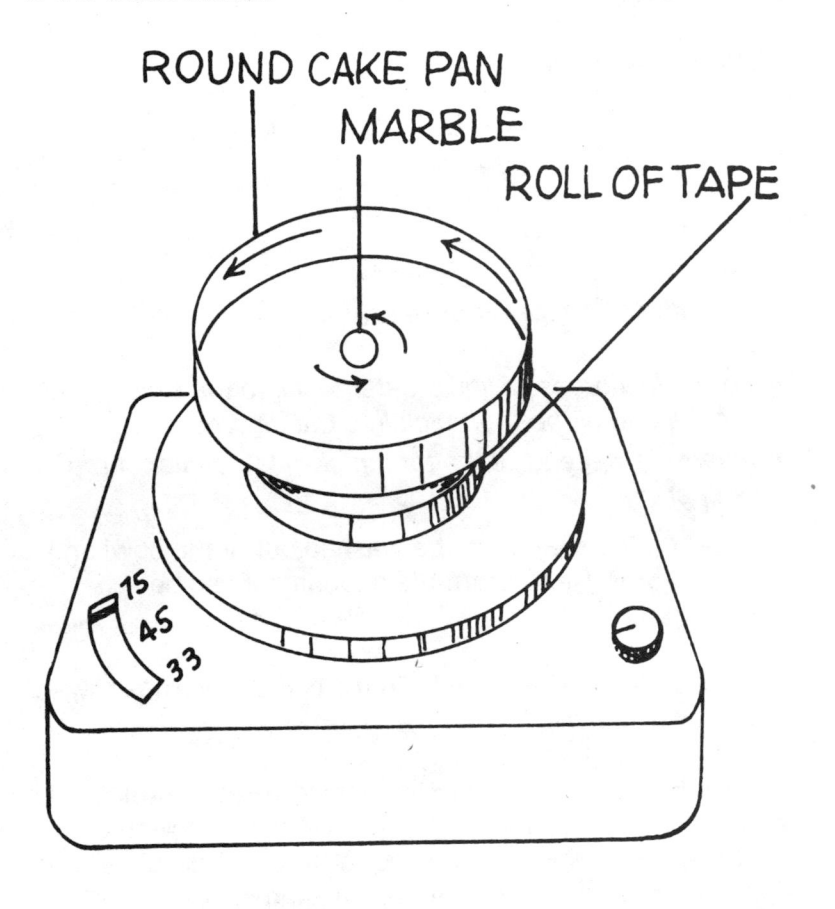

ROUND CAKE PAN
MARBLE
ROLL OF TAPE

74. Pepper Run

Purpose To make black pepper run across a bowl of water with soap as its power source.

Materials *black pepper*
toothpicks
bowl full of water, 2 qt. (2 liter)
liquid detergent
saucer

Procedure

- *Sprinkle the pepper over the surface of the water in the bowl.*
- *Pour a few drops of liquid detergent into the saucer and dip the end of the toothpick into the detergent.*
- *Insert the wet end of the toothpick in the center of the pepper.*

Note: All the soap must be washed out of the bowl and fresh water used before the experiment can be repeated.

Results The pepper breaks in the center and runs toward the sides of the bowl.

Why? Each pepper speck is waging a tug-of-war battle. While the water is clean, the surface water molecules pull on the pepper specks with equal force in all directions. Placing the soap in the center weakened the pull of the water molecules in that area, and the stronger, clean-water molecules pulled the pepper specks across the surface of the water toward the side of the bowl.

170

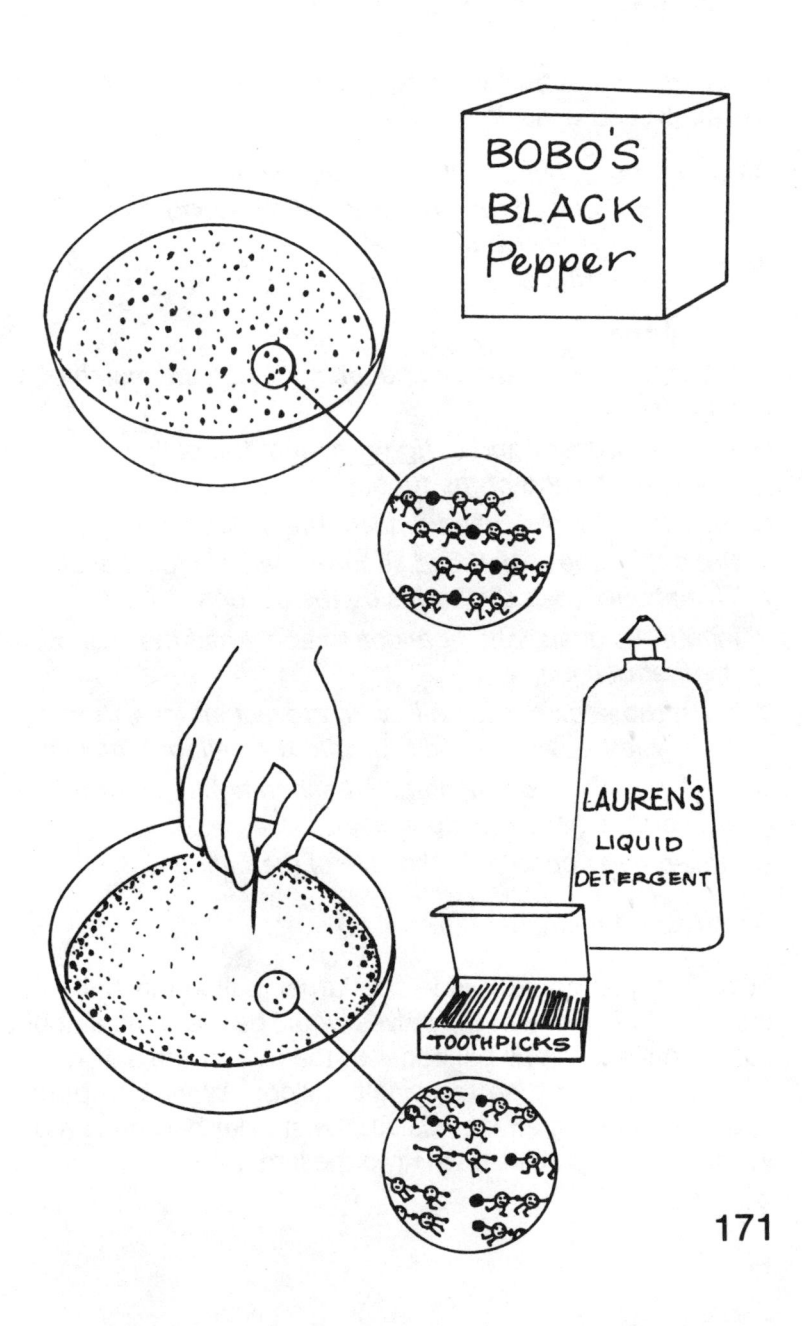

171

75. Which Way

Purpose To determine how the size of a balloon affects the air pressure inside it.

Materials 2 round balloons, 9 in. (23 cm)
plastic tubing, 12 in. (30 cm) long
rubber band
clothespin

Procedure
- Insert one end of the plastic tubing inside the mouth of a balloon.
- Use the rubber band to tightly secure the balloon's mouth to the end of the tube.
- Blow through the tube to inflate the balloon.
- Bend the tube in half and secure it with the clothespin. This should keep the air inside the balloon.
- Inflate the unattached balloon to about half the size of the first balloon.
- Twist the neck of the balloon to prevent air loss, then insert the free end of the tube into the balloon's mouth.
- Hold the mouth of the smaller balloon tightly around the tube and remove the clothespin.
- Observe the change in the size of the balloons.

Results The smaller balloon deflates.

Why? If you could divide the larger balloon in half, the amount of air in one of the halves would be the same amount as in the single small balloon, but the size of the half would be much smaller than the single balloon. When comparing equal amounts of air molecules, the smaller balloon is more stretched and pushes its air into the larger balloon.

172

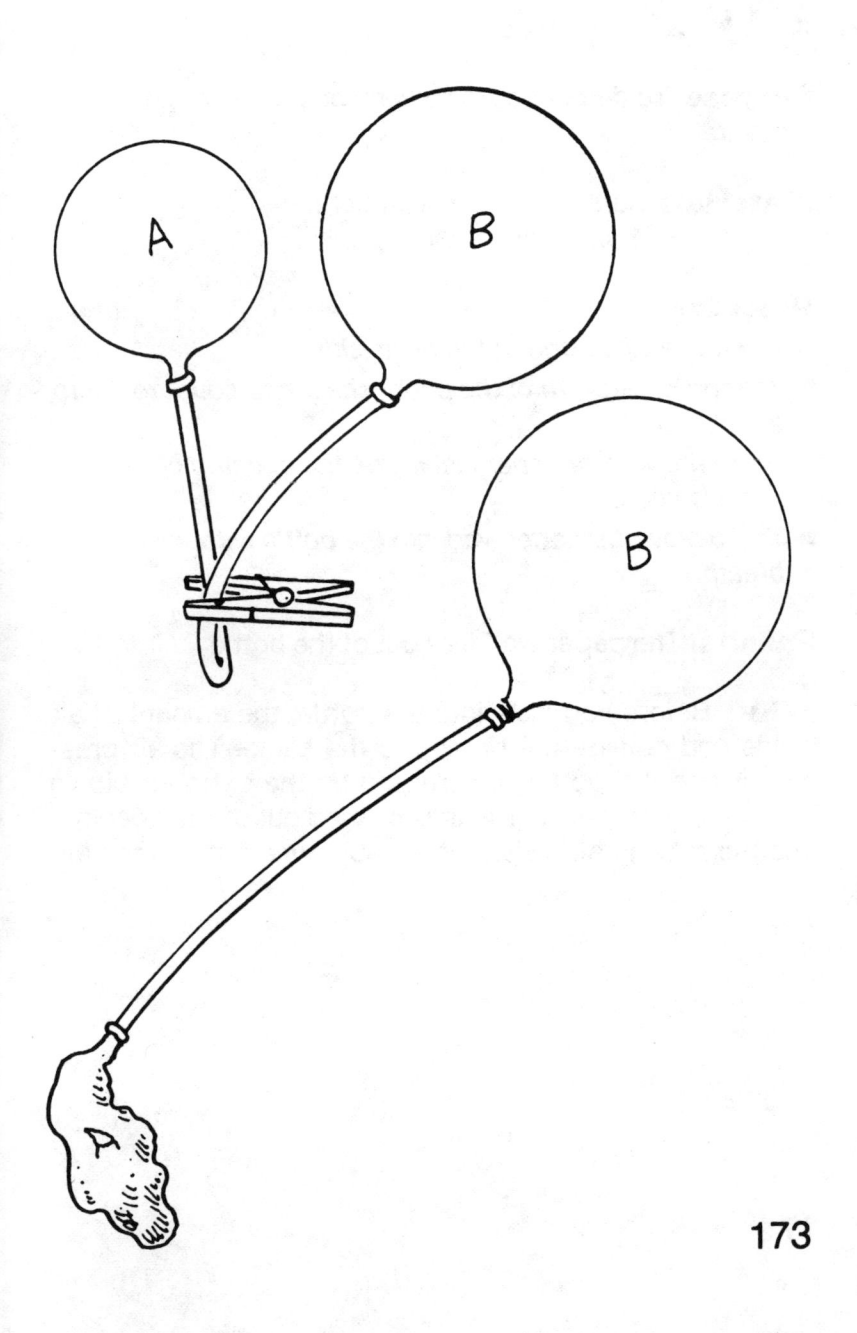

76. Fly Away

Purpose To demonstrate movement of air due to unequal pressure.

Materials *narrow-necked soda bottle*
1 sheet of notebook paper

Procedure
- *Lay the empty soda bottle on its side.*
- *Tear off one-fourth of the paper sheet and squeeze it into a ball.*
- *Place the wad of paper just inside the opening of the bottle's neck.*
- *Try to blow the paper wad into the bottle with your breath.*

Results The paper wad flies out of the bottle.

Why? Before you blow into the bottle, the amount of air inside and outside the bottle are the same. The air pressure inside the bottle is increased by the extra air blown into the bottle. This extra air is pushed out of the opening, and the moving air swishes the paper wad out into the air.

X
Light

77. Waves

Purpose To demonstrate how sound and light travel.

Materials *Slinky®, about 4-in. (10-cm) long*
helper

Procedure
- *Have your helper hold one end of the Slinky and stretch it about four times its length while you hold the other end.*
- *Shake your end of the Slinky up and down several times.*
- *Notice how the Slinky moves.*
- *Place the Slinky on the floor and stretch it as far as possible.*
- *Reach as far as you can down the stretched spring and gather the coils toward you, then quickly release them.*
- *Notice how the Slinky moves.*

Results Shaking the Slinky up and down makes it look like water waves. Compressing the coils sends a back-and-forth pulsing movement down the spring.

Why? The compression-and-expansion type of motion is called a *longitudinal wave. Sound* is produced by a vibrating material that moves back and forth. As the material moves outward, it pushes the air in front of it together, producing a compressed area of air molecules. When the material swings back, it leaves an expanded area. This back-and-forth motion produces compressed and expanded area in the air through which the energy is being transferred. This is how sound vibrations move through the air.

The up-and-down movement of the Slinky represents a *transverse wave.* These are like water waves in that they have high and low parts called crests and troughs. Light travels in transverse waves.

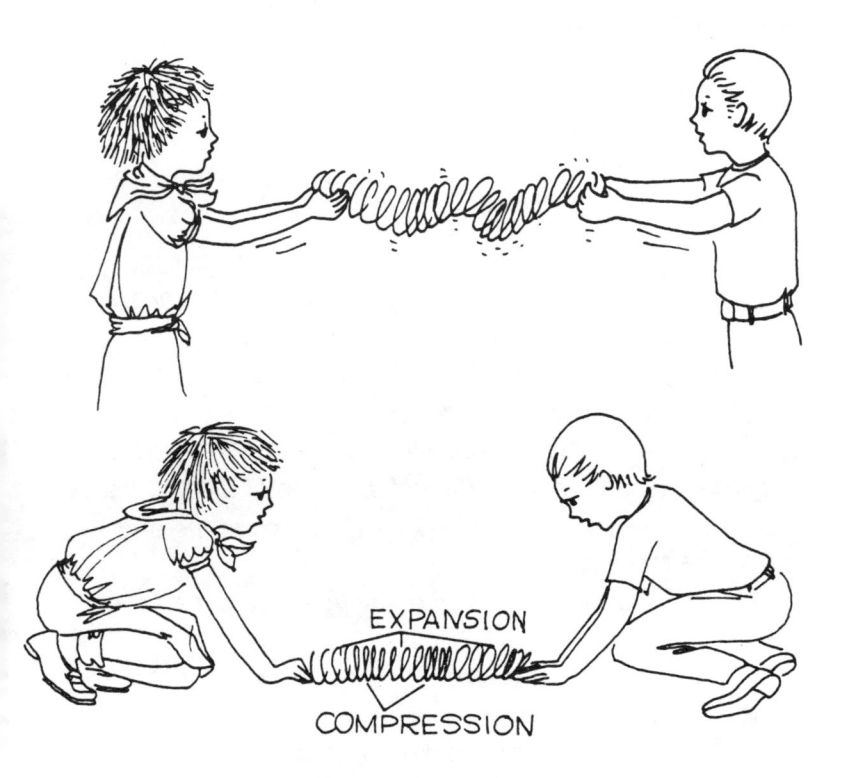

EXPANSION

COMPRESSION

78. Straight

Purpose To demonstrate that light travels in a straight line.

Materials *cardboard*
flashlight
scissors
modeling clay
ruler
index card

Procedure

■ *Cut three 6-in. (15-cm) squares from the cardboard.*

■ *Cut 1-in. (2.5-cm) square notches from the center of one edge of each of the three cardboard squares.*

■ *Use the clay to position the cardboard about 4 in. (10 cm) apart with the notches aligned in a straight line.*

■ *Lay the flashlight behind the notch of one of the end cards.*

■ *Use clay to position the index card screen at the end opposite the flashlight.*

■ *Darken the room and observe any light pattern on the paper screen.*

■ *Move the cardboard so that the notches are not in a straight line.*

■ *Observe any light pattern on the paper screen.*

Results A circle of light appears on the screen only when the notches are in a straight line with each other.

Why? Light travels in a straight line. When the notches were in line, the light rays were able to pass through the openings, but the rays were blocked by the cardboard when the notches were out of line.

180

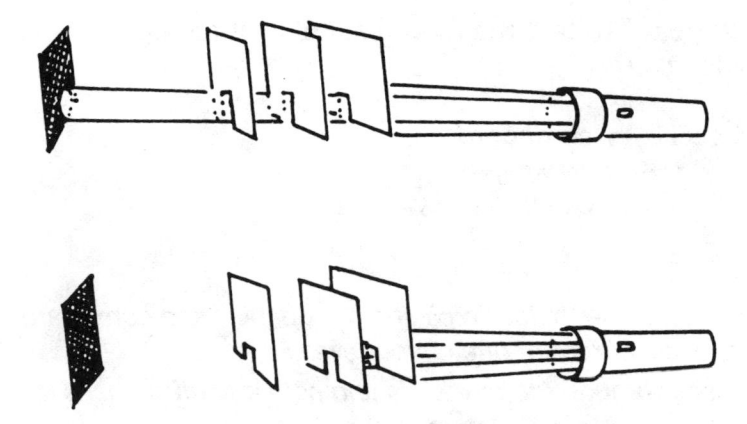

79. See Through

Purpose To test the movement of light through different materials.

Materials *cardboard*
wax paper
plastic sandwich bag

Procedure
- *One by one, hold the plastic, wax paper, and cardboard pieces in front of one of your eyes.*
- *Look through the materials and note any differences in how objects in the room appear.*

Results There is little or no change in appearance when things are observed through the plastic. The wax paper makes objects look dull and frosty, while nothing can be seen through the cardboard.

Why? In order for you to see anything, light must be reflected from the object you are looking at to your eyes. The clear plastic is an example of a *transparent* material. Transparent means that light moves straight through the material and allows you to see objects as they are. *Translucent* materials, like wax paper, change the direction of the light that passes through. This change in direction results in objects looking dull, frosty, and sometimes distorted. Cardboard is an *opaque* material—no light can pass through. Without light passing through to your eye, nothing on the opposite side of opaque materials can be seen.

183

80. Polarized Light

Purpose To determine how polarized light moves.

Materials *two pairs of polarized sunglasses*
Note: Be sure you are using polarized glasses.

Procedure
- *Put one pair of glasses on.*
- *Observe how objects around you appear.*
- *Hold the second pair of glasses in front of your eyes.*
- *Slowly rotate the pair of glasses you are holding so that one of the lens turns in front of your right eye.*
- *Observe the quality of vision as you turn the glasses.*

Results One pair of sunglasses seems to cut down glare and change the shade of objects. During the rotation of the second pair of glasses, the objects seen through the right eye got darker until finally nothing could be seen through the two lenses.

Why? A polarized lens has an endless number of parallel slits. Light waves moving in the same direction as the slits are allowed to pass through. Light not moving in the same direction as the slits in the lens is blocked and cannot pass through.

184

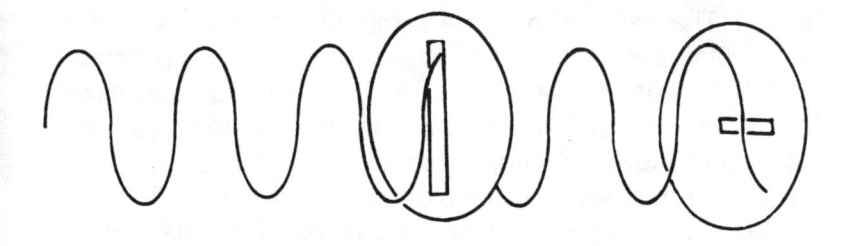

81. Swirls of Color

Purpose To separate light into colors.

Materials bowl of water, 1 qt. (1 liter)
bottle of clear fingernail polish

Procedure
- Place the bowl of water on a table away from direct lighting.
- Hold the brush from the polish bottle over the bowl of water and allow one drop of liquid polish to fall into the bowl.
- Watch the surface of the water. Move your head so that you see the surface from different angles.

Results A rainbow of colors is seen in the thin layer of fingernail polish on the surface of the water.

Why? The nail polish forms a thin film across the water. When light rays strike the film, part of each ray is reflected from the surface. Part of the ray goes through and is reflected off the bottom of the film. If the reflected rays overlap as they leave the film, colors are seen. The thickness of the film determines the speed of the reflected rays. The timing has to be just right for the reflected rays from the surface and bottom of the film to meet as they leave the film; if this does not happen, areas without colors are seen. This rainbow of colors is called a *spectrum.*

186

82. Water Prism

Purpose To use water to separate light into its separate colors.

Materials *flashlight*
heavy paper
masking tape
scissors
sheet of white paper
chair
glass of water
helper

Procedure

■ *Cut a circle from the heavy paper to cover the end of the flashlight.*

■ *Cut a very thin slit across the circle, stopping about 1¹/₂ in. (1 cm) from the edge.*

■ *Tape the paper circle to the front of the flashlight.*

■ *Place the glass of water on the edge of the chair.*

■ *Have your helper hold the white paper near the floor at the edge of the chair.*

■ *Darken the room and hold the flashlight at an angle to the surface of the water.*

■ *Change the angle of the flashlight and ask your helper to vary the position of the white paper.*

■ *Look for colors on the white paper.*

Results A *spectrum* of colors is seen on the paper.

Why? White light contains all of the spectrum colors—red, orange, yellow, green, blue, indigo, and violet. Light can be separated into the spectrum colors by passing it through different substances such as water or glass. Light must be *refracted*—bent or spread out—in order to be split into a spectrum. Scientists use this property of refraction to determine the density and composition of materials through which light can pass.

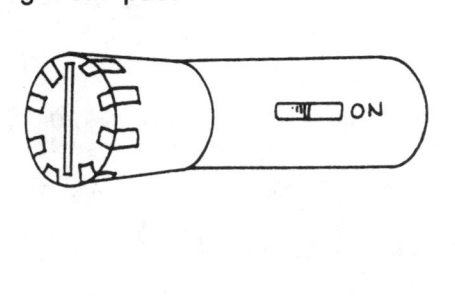

83. Blender

Purpose To demonstrate how light waves blend to produce white light.

Materials *poster board*
scissors
pencil
ruler
marking pens, red, orange, yellow, blue, green, and violet

Procedure

- *Cut a 4-in. (10-cm) circle from the poster board.*
- *Divide the circle into 6 equal sections.*
- *Color the sections in this order: red, orange, yellow, blue, green, and violet.*
- *Punch the pencil through the center of the circle, leaving about half of the pencil on each side.*
- *Place the point of the pencil on a flat surface and quickly spin the top of the pencil.*
- *Observe the color of the spinning circle.*

Results The spinning circle looks grey.

Why? White light is made up of all the spectrum colors—red, orange, yellow, green, blue, indigo, and violet.

As the colored circle rotates, the different colors are perceived by your eyes. The images of the colors are retained in your mind long enough to cause an overlapping of the messages. Your own mind blends the colors together. If the color indigo—a dark bluish color—were included, you would see a twirling white card because all of the spectrum colors blended together produce white light.

190

84. Rosy Skin

Purpose To determine how filters affect light.

Materials *flashlight*

Procedure
- *Darken a room and hold the flashlight under your hand.*
- *Move the light around behind your fingers and palm.*
- *Observe any light that passes through.*

Results Parts of your hand appear rosy in color.

Why? Your flesh and skin act like filters. A filter is any material that absorbs some of the colors in light and allows others to pass through. Red filters absorb all colors except red which passes through. Your skin takes on a rosy color because the red blood under the skin acts like a red filter—red light passes through and other colors are stopped.

85. Backwards

Purpose To determine how a mirror affects the reflected image.

Materials *hand mirror*
pencil
paper
4 books

Procedure

- *Support the mirror with the books.*
- *Place the paper under the edge of the mirror.*
- *Rest your chin on your hand so that you can see into the mirror, but so that your view of the paper where you will be writing is blocked.*
- *Look only into the mirror as you write your name so that it appears correctly in the mirror.*
- *Examine your writing.*

Results Most and maybe all of the letters are upside down.

Why? Because you are writing so that the letters are correct in the mirror, they are reversed on the paper. Most letters would be written upside down for this experiment except those that are symmetrical (the same on both sides) such as O, E, H, I, and B. These letters look the same with or without the mirror. The mirror gives you a reversed image.

86. Blinker

Purpose To demonstrate the strobe effect of a television picture.

Materials *television*
  pencil

Procedure

- *With the lights on, swing the pencil up and down quickly and observe how it looks.*
- *In a darkened room with only the television on, hold the pencil in front of the screen.*
- *Quickly swing the pencil up and down 4 to 5 times.*

Caution: Never watch television in a room lit only by light from the television screen. It is extremely bad for your eyesight.

Results In light, the moving pencil produces a continuous blur. In the dark, separate images of the pencil are seen in different places.

Why? Instead of a continuous blur of the moving pencil, separate images are seen, because the light from the television screen is not constant. Thirty pictures are flashed on the screen each second. Between the pictures the screen goes black, and the pencil moves to a new place. The movement of the pencil is not seen when there is no light. This blinking light gives the illusion that the pencil is moving in slow motion. You do not notice any blinking while watching television because your eyes retain the image of each picture long enough to receive the flashing of the next picture.

196

87. Bent

Purpose To demonstrate how a diffracting grating affects light.

Materials *cotton handkerchief*
lamp

Procedure
- *Remove the shade from the lamp.*
- *Stand about 6 ft. (2 m) from the glowing bulb.*
- *Look at the light through the stretched cloth.*

Results A starburst of light with dim bands of yellow and orange colors appears around the light.

Why? The cloth acts like a *diffracting grating,* which separates light into its individual colors. Diffracting gratings are made by using a diamond point to cut as many as 12,000 lines per ½ in. (centimeter) on a piece of glass or plastic. The spaces between the woven threads in cloth separate the light, but since the holes in the weave are large, not as many separate colors are seen as one would observe through a diamond cut grating. The starburst pattern is due to the bending or changing direction of the light.

XI

Heat

88. Chilling

Purpose To determine if gas pressure affects temperature.

Materials *aerosol can of first-aid medicine*

Procedure
- *Hold the spray can near your arm.*
- *Spray some of the contents onto your arm and note the temperature of the spray.*

Results The spray feels cold.

Why? The can contains medicine and gas to produce a high pressure. When the nozzle opens, the gas escapes quickly, carrying with it drops of liquid. The cold temperature is due to the rapid expansion of the gas. Increasing the volume of the gas decreases the pressure on the gas and causes its temperature to drop.

89. Hot Band

Purpose To demonstrate energy changes.

Materials *rubber band*

Procedure

- *Place the rubber band on your forehead and note the rubber band's temperature.*

  Note: Your forehead is sensitive to heat and can therefore be your sensing device.
- *Hold the rubber band between your thumbs and index fingers with your thumbs touching.*
- *Stretch the rubber band.*
- *Quickly touch the stretched band to your forehead.*

Results The stretched rubber band feels warm.

Why? The rubber band is made of molecules coiled like a spring. Stretching the rubber band straightens the coils; they recoil when the band is released. You used *mechanical energy* —energy of moving things—to pull the coils of molecules apart, and the rubber band used energy to pull the coils back together. Some of the mechanical energy was changed into *heat energy.* Energy was needed to stretch the rubber band, and energy was needed to restore it to its original shape. If there were no changes in the molecular structure of the rubber band, the amounts of energy used to stretch and to recoil the rubber band would be the same. The energy changed from one form to another, but it was not lost. This is called *conservation of energy.*

205

90. Cold Foot

Purpose To demonstrate the conduction of heat energy.

Materials *aluminum foil*
small throw rug or rug sample

Procedure
- *Cut a piece of foil a little bigger than your foot.*
- *Place the foil and the carpet on a tile floor. Allow them to remain undisturbed for 10 minutes.*
- *Put one bare foot on the aluminum and the other on the carpet.*
- *Observe any difference between the feel of the temperature of the aluminum piece and that of the carpet.*

Results The metal foil feels colder than the carpet.

Why? A good heat conductor allows heat to move through it. This is accomplished by movement of individual molecules. One molecule receives the heat energy and starts bouncing around. During these jerky motions, the molecule bumps into a molecule nearby, causing it to bounce around. This exchange of energy continues through the entire piece of material if it is a conductor of heat. Some materials are very good conductors of heat and others are not. Molecules in metals like aluminum can easily move around and are very good conductors of heat energy. Good heat conductors are also good electrical conductors. The piece of carpet is a poor conductor of heat, which means that its molecules do not move easily. Allowing the foil and carpet to sit for 10 minutes gave both materials time to reach room temperature.

Things feel cold to the touch when heat energy is drawn away from your skin; things feel warm when heat energy is transferred to your skin. The aluminum feels colder than the carpet because it is a good heat conductor and the heat from your foot starts moving through the metal. The carpet is a poor heat conductor and actually blocks heat loss from your foot.

Poor heat conductors are also called *insulators.* To keep a house from losing its energy, insulators are placed in the space between the inside and outside walls to block the flow of heat into or out of the house.

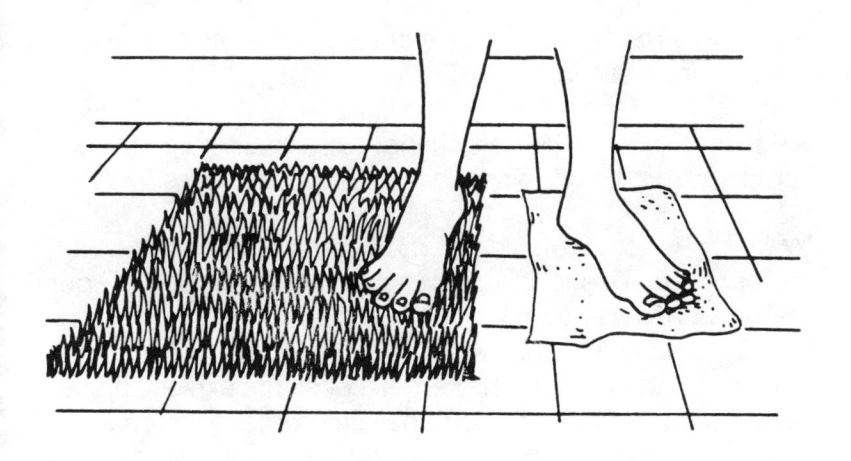

91. Explosive

Purpose To determine why popcorn pops.

Materials *unpopped popcorn*
hot-air popcorn popper

Procedure
- *Observe the shape and size of a few unpopped corn kernels.*
- *Ask an adult helper to assist you in setting up the popcorn popper.*
- *Observe the shape and size of the corn kernels as they are being heated.*

Results The corn kernels change from small, hard, orange, kernels to large, soft, white, ball-shaped structures.

Why? The tough outside of the unpopped kernel is called the *pericarp*. This is the part that often gets stuck in your teeth when you eat popcorn. The inside is filled with starch that expands into the white, fluffy popcorn. The small amount of water inside the kernel makes the explosion possible. As the kernel is heated, the liquid water *evaporates*—changes to a gas. The gas expands and pushes so hard on the pericarp that it breaks and the starch tissue inside is blown outward. The pop noise is the sound of steam escaping and the pericarp breaking.

92. Bouncer

Purpose To determine if temperature affects the bounce of a rubber ball.

Materials *tennis ball*
yardstick (meter stick)
refrigerator with freezer

Procedure
- *Hold the yardstick with one hand and place the ball at the top edge of the yardstick.*
- *Release the ball, and observe the height of the first bounce. Repeat three times to get an average of the bouncing height.*
- *Place the ball in a freezer for 30 minutes.*
- *Again measure the height that the ball bounces when released from the top of the yardstick.*

Results The ball does not bounce as high when it is cold.

Why? Rubber is made of thousands of small molecules joined to form long chains. At room temperature, the chains easily push together and pull apart to make the ball bounce. The chains of molecules become rigid when the ball is chilled. The flexibility of the chains of molecules allows the tennis ball to bounce. Playing tennis in cold weather would affect your game.

XII

Sound

93. Boom!

Purpose To demonstrate the effect of solids on the speed of sound.

Materials *clear plastic drinking glass*
rubber band

Procedure
- *Stretch the rubber band around the glass as shown in the diagram.*
- *Hold the bottom of the glass against your ear.*
- *Gently strum the stretched rubber band.*

Results The sound heard is very loud.

Why? Sound is produced when objects vibrate. As the object moves back and forth, it hits against air and any other object near enough to be touched. When vibrations start air moving, the continuous ocean of air around you transfers the energy to your ears, and you register that sound has been produced. Vibrations move much more slowly through the air—a gas—than they do through liquids or solids. The vibrating rubber band causes the air around it to move, but the booming sound that you hear is because the solid plastic transmits the vibrations to your ear. You receive many more vibrations in a short time period through the solid than through air.

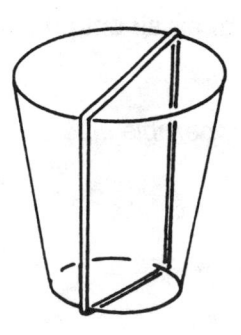

94. Bottle Organ

Purpose To demonstrate how the height of an air column affects the pitch of sound.

Materials *6 small-mouthed bottles of comparable size*
metal spoon

Procedure
- *Pour different amounts of water in each bottle.*
- *Gently tap each bottle with the metal spoon.*
- *Note the difference in the pitch produced.*

Results The bottle with the most water has the highest pitch.

Why? Sounds are made by vibrating objects. The number of times the object vibrates—moves back and forth—is called the *frequency* of the sound. As the frequency increases, the *pitch* of the sound gets higher. Tapping on the bottles causes the air in the bottles to vibrate. The shorter column of air above the water vibrates faster, producing a higher pitch. As the height of the air column increases, the pitch of the sound gets lower.

95. Singing Glass

Purpose To show how friction can cause a glass to vibrate.

Materials *stemmed glassware (this will work better if the glass is very thin)*
dish detergent
sink or large pan
vinegar
small shallow bowl

Procedure

- *Use the dish detergent to make a warm, soapy water solution in a sink or large pan.*
- *Wash the glass and your hands in the warm, soapy water, and rinse well.*
- *Place the glass on a table.*
- *Pour a thin layer of vinegar into the small bowl.*
- *Hold the base of the glass against the table with your hand.*
- *Wet the index finger of your free hand with vinegar and gently rub your wet finger around the rim of the glass.*

Results The glass starts to sing when its rim is rubbed.

Why? Washing the glass and your hands removes any oil that might act as a lubricant. The vinegar also dissolves any oil that might be present and increases the friction between your skin and the glass. Rubbing your finger around the rim causes the glass to vibrate because your finger skips and pulls at the glass. This irregular touching on the glass rim is

218

like tiny taps that cause the glass to vibrate. The air inside the glass is struck by the vibrating glass and begins to move back and forth in rhythm with the glass. The pitch of the sound you hear is due to the number of air molecules that strike your eardrum.

96. String Music

Purpose To demonstrate factors that affect the pitch of stringed instruments.

Materials *2 buckets, 1 qt. (1 liter) is large enough*
rocks, enough to fill both buckets
2 pencils
string
scissors
table, about 3 ft. (1 m) wide

Procedure

■ *Cut a length of string twice as wide as the table.*
■ *Place the string across the table and tie the ends to the bucket handles so that the buckets hang freely.*
■ *Place the pencils on the edge of the table under the string.*
■ *Fill each bucket half full with rocks.*
■ *Pluck the center of the string with your fingers.*
■ *Move the pencils closer together and again pluck the center of the string.*
■ *Fill the buckets with rocks, move the pencils to different positions, and pluck the string at each different pencil position.*

Results Adding rocks to the buckets and moving the pencils closer together produces a higher pitched sound when the string is plucked.

Why? The *pitch* of a vibrating string obeys certain laws. Length and tension are two factors that affect the pitch of sound produced by a stringed instrument. The pitch is a

220

result of frequency (the number of vibrations produced per second). The faster the string vibrates, the higher is the pitch of the sound produced. The string vibrated faster as the tightness of the string or tension was increased by adding weight to the buckets. Moving the pencils closer together shortened the string that vibrated. The shorter the string, the faster it vibrated. Increasing the tension and shortening the length of a plucked string produces a higher pitched sound because the string vibrates faster. This is what happens when a guitar string is held down with one hand while being plucked with the other. Violins, cellos, and other instruments with strings work the same way.

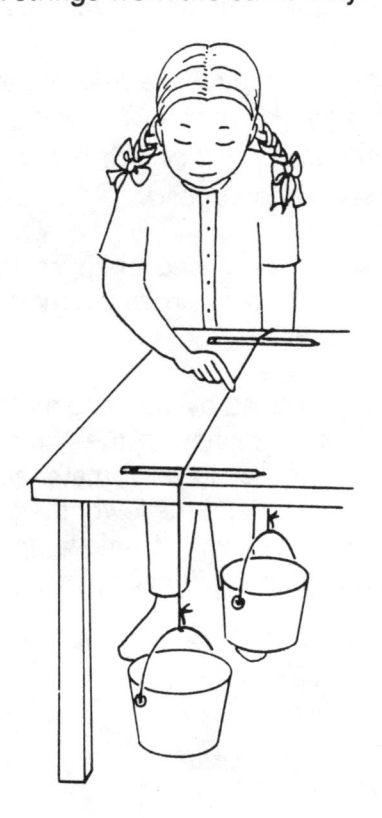

97. Twang

Purpose To demonstrate the effect that length has on the sound of a vibrating material.

Materials *ruler*
table

Procedure
- *Place the ruler on a table with about 10 in. (25 cm) of the ruler hanging over the edge of the table.*
- *Press the end of the ruler against the table with your hand.*
- *With your other hand, push the free end of the ruler down and then quickly release it.*
- *As the ruler moves, slide it quickly onto the table.*
- *Observe the sounds produced.*

Results The sound produced changes from a low to a high pitch as the length of the ruler extending over the edge of the table decreases.

Why? Sound is produced by vibrating materials. The *pitch* of the sound becomes higher as the number of vibrations increases. The longer the vibrating material, the slower the up-and-down movement and the lower the sound produced. Shortening the ruler causes it to move up and down very quickly, producing a higher pitched sound.

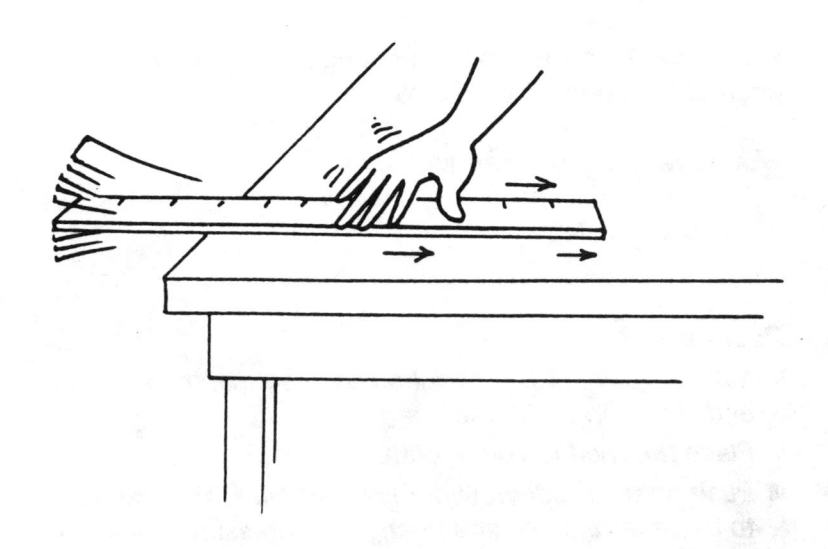

98. Straw Flute

Purpose To determine if the length of a flute affects the pitch of the sound it produces.

Materials *drinking straw*
scissors
ruler
marking pen

Procedure
- *Make a ¹/₂-in. (1.3-cm) cut on each side of the straw's end. This forms the reed part of the flute.*
- *Place the reed in your mouth.*
- *Push on the reed with your lips and blow. You may have to try several times and change the pressure of your lips in order to produce a sound.*
- *As you play the straw flute, cut the end of the straw off with the scissors and observe any change in pitch.*

Results The *pitch* of the sound gets higher as the length of the straw decreases.

Why? The sound produced is due to the vibration of the straw and the air inside it. The longer the column of vibrating air inside the tube, the lower is the pitch of the sound.

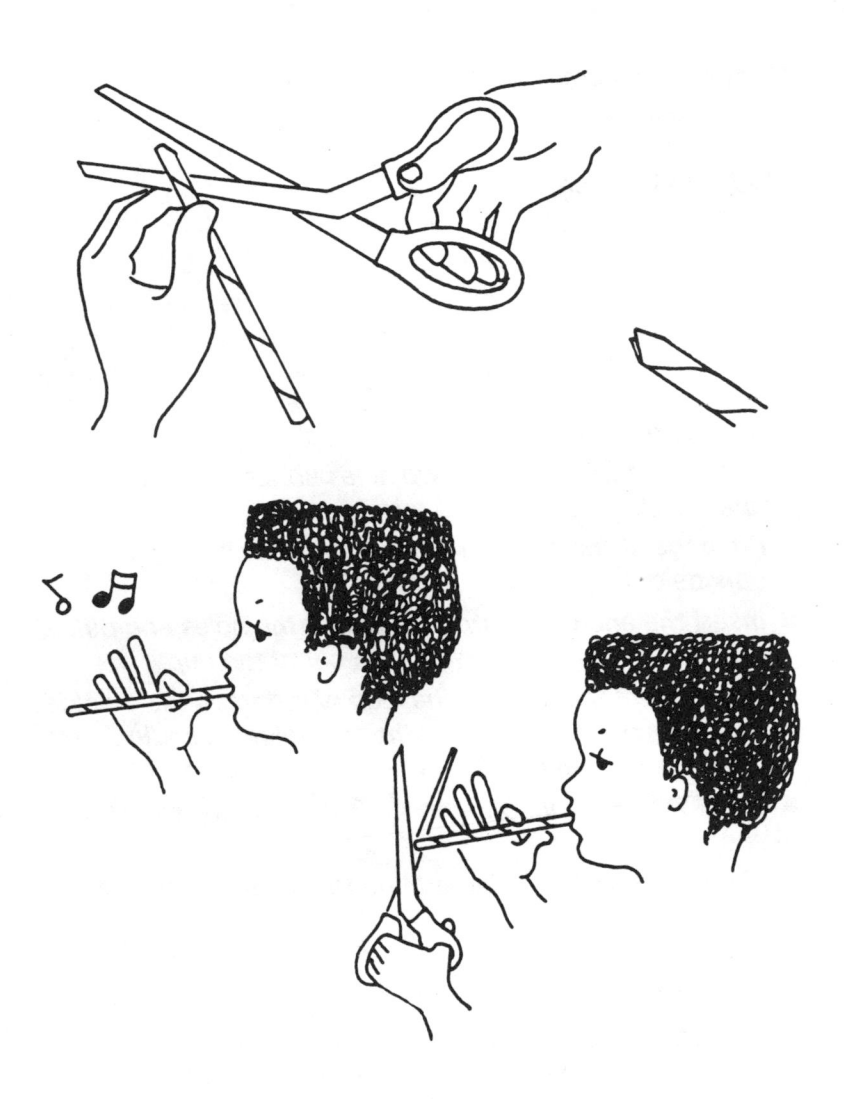

225

99. Clucking Chicken

Purpose To use a vibrating string to produce a clucking chicken sound.

Materials *paper cup, 6.4 oz. (192 ml)*
kite string, 24 in. (60 cm)
sponge, rectangular kitchen sponge
pencil
toothpick
water

Procedure

- *Use the pencil to punch two holes about 1/2 in. (1.5 cm) apart in the bottom of the cup.*
- *Push the string through the holes and tie it on the outside of the cup.*
- *Insert the end of the string in one of the holes and pull it through so that the string hangs out of the cup.*
- *Place a toothpick under the loop of string on the outside of the cup with the ends of the toothpick extending over the edges of the cup.*
- *Cut a 1-in. × 1/2-in. (2.5-cm × 1.3-cm) section from the sponge.*
- *Tie the end of the string around the center of the piece of sponge.*
- *Wet the sponge with water.*
- *Wrap the wet sponge around the top of the string.*
- *Squeeze the sponge against the string as you move the sponge down the string using jerky movements.*

Results A sound is produced like that of a clucking chicken.

Why? The water allows the sponge to move down the string, but there is enough friction to cause the string to vibrate, because the sponge skips and pulls at the string. This irregular touching on the string produces tiny taps that force the string's molecules to move back and forth. The vibrating string strikes the molecules in the cup, and the cup's molecules strike the air molecules, causing them to move back and forth in rhythm with the cup and string. The *pitch* of the sound is due to the number of air molecules that strike your eardrum. The sound is made louder because the inside of the cup acts like a megaphone that concentrates the sound waves and sends them out in one direction.

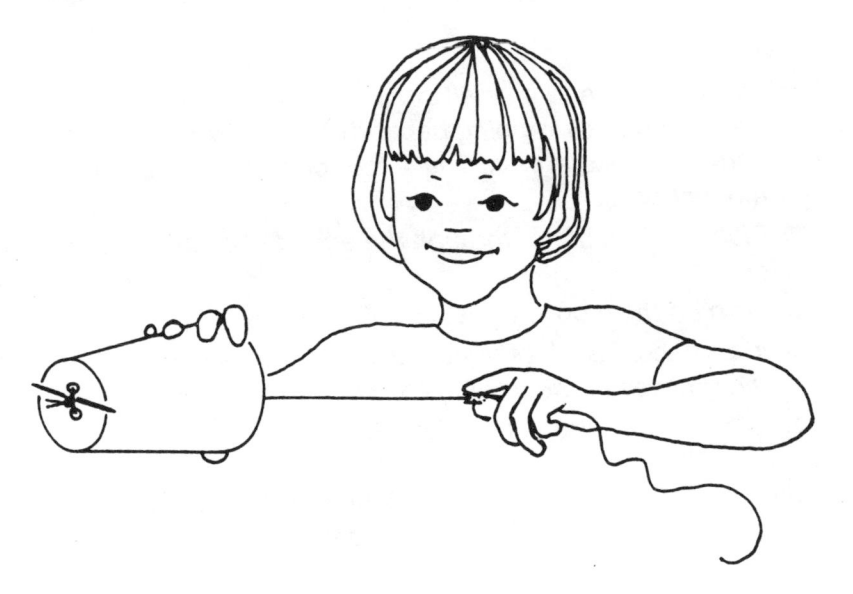

100. Phonograph Speaker

Purpose To determine the best structure of a phonograph speaker.

Materials *poster board*
wrapping paper
33-RPM record—one that is disposable!
scissors
masking tape
4 straight pins
record player

Procedure
- *Make a large and a small cone from both the poster board and the wrapping paper by rolling and taping as shown in the diagram. These four cones will be used as megaphones.*
- *Place a straight pin through the end of each of the four cones.*
- *Turn on the record player.*
- *In turn, hold each of the cones so that the pin rests on the grooves of the turning record.*
- *Observe any sounds produced.*

Results Music is heard. The loudest and clearest sound was heard through the megaphone made of large, thin paper.

Why? Sound is caused by vibrating air that reaches your ear. The pin moves up and down in the record grooves and hits against the paper megaphone. The megaphone vibrates the air inside it. It is this moving air that reaches your ear. The thinner paper works best because it is more easily moved. The larger cone gives more paper to vibrate and thus increases the volume of the sound.

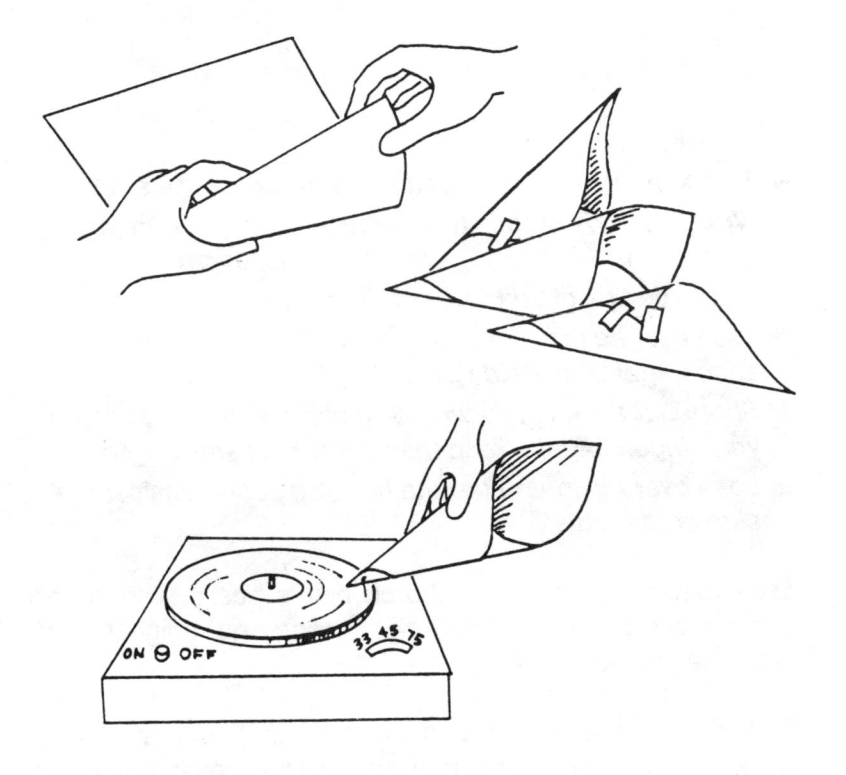

101. Spoon Bell

Purpose To demonstrate how the pitch of sound can be changed.

Materials *metal spoon*
kite string, 30 in. (75 cm)

Procedure

- *Tie the handle of the spoon in the center of the string.*
- *Wrap the ends of the string around your index fingers. Be sure that both strings are the same length.*
- *Place the tips of your index fingers in each ear.*
- *Lean over so that the spoon hangs freely and tap the spoon against the side of a table.*
- *Shorten the strings by wrapping more of the string around your fingers. Again, keep the strings the same length.*
- *Lean over again and tap the hanging spoon against the side of the table.*

Results A sound like a church bell is heard. The sound has a lower *pitch* with the longer string and a higher *pitch* with the shorter string.

Why? The metal in the spoon starts to vibrate when struck. Vibration means that something moves back and forth. In this case, the molecules in the spoon move back and forth and hit against each other. When molecules collide, energy is transferred from one molecule to the other. The vibrating molecules in the spoon hit against the string's molecules, and the energy is transferred up the string to your ears. The difference in the pitch of the sound is due to frequency (the number of vibrations that reach the ear in one second). High-pitched sounds are produced when many vibrations reach the ear in one second, and the sound gets lower in pitch as the number of vibrations decreases.

The long and the short strings both receive about the same amount of energy from the vibrating spoon. Since the shorter string has fewer molecules, each molecule receives a bigger share of the energy and vibrates faster, thus producing a higher pitched sound. The long string produces a lower pitch because there are more molecules to share the energy, causing each molecule to be less energetic. This results in a slower vibration.

Glossary

Air Resistance Upward force that air exerts on falling objects. It is also called air friction.

Attract To be pulled together.

Atom The smallest part of an element that contains a positive center with negative charges spinning around the outside.

Buoyant Force The upward force that a liquid exerts on an object. The force is equal to the weight of the liquid that is pushed aside when the object enters the liquid.

Center of Gravity The point where an object will balance. The weight of the object is evenly spaced around this point.

Conductor A material through which electrons move freely. Metals are good conductors.

Drag Air resistance to forward motion.

Electric Circuit The path through which electrons (negatively charged atomic particles) move.

Electrical Impulses Transfer of energy from one electron to the next through a metal wire.

Electroscope An instrument that detects static charges.

Electrons Negatively charged particles spinning around the atom's nucleus (the center part of every atom).

Filter A material that absorbs some of the colors of light and allows others to pass through.

Fluorescence A form of luminescence (giving off of light) by changing invisible ultraviolet light into visible light.

Friction A force that pushes against a moving object, causing it to stop moving.

Fulcrum The point of rotation on a lever.

Gravity The downward pull on all objects, referred to as the weight of the object.

Hypothesis An educated guess about what you think the solution to a problem is.

Inclined Plane A simple machine that has a tilted surface. Used to move objects to a higher level.

Inertia Inertia is the resistance that an object has to having its motion changed. Objects that are stationary continue to be at rest, while objects in motion continue to move due to their inertia. Stationary objects move and moving objects stop due to some outside force that pushes on them.

Insulator A material that blocks the flow of electrons.

Kinetic Energy Energy of motion having magnitude as well as direction.

Law of Conservation of Energy Energy is never created or lost, just transferred or changed into another form.

Lift The upward force on flying craft due to the difference in the speed of air flowing over and under the wings.

Longitudinal Wave A wave that has a back-and-forth motion.

234

Machines Make work easier, faster, can change the direction of the force applied, and can increase the input force.

Magnetic Domains Clusters of atoms that behave like tiny magnets.

Magnetic Field The area around a magnet in which the force of the magnet affects the movement of other magnetic objects.

Magnetic Pole Magnets have a north and south pole. The north end of the magnet points to the earth's north magnetic pole.

Mass The amount of matter in an object.

Matter Anything that takes up space and has weight. Matter is composed of atoms.

Molecule Two or more atoms joined together.

Newton's Law of Action-Reaction When an object is pushed, it pushes back with an equal and opposite force.

Nucleus The center of all atoms, containing positively charged protons and neutral neutrons, which gives it an overall positive charge.

Opaque A material that light cannot pass through.

Pericarp The tough outer portion of a kernel of corn.

Phase of Matter The three phases most often seen are gases, liquids, and solids.

Pitch The number of vibrations that reach the ear in one second. The pitch of the sound increases as the vibrations increase.

Polarized Lens Allows light rays moving in one direction to pass through.

Potential Energy Stored energy due to position. The higher the object, the more potential energy it has.

Protons Positive particles in the nucleus of all atoms.

Repel To move away from. Like charges repel or move away from each other.

Reflect To bounce back from a surface.

Refract When light goes from one substance to another it changes direction. Sometimes the change of direction produces a spreading out of the light rays and a spectrum is produced.

Sound Vibrating energy that causes the sensation of hearing.

Spectrum The colors found in white light—red, orange, yellow, green, blue, indigo, and violet.

Static Electricity A build-up of negative charges called electrons. Static means stationary.

Switch Any material, usually metal, that acts like a draw bridge in an electric circuit. When it is closed, the electrons continue to travel, but when it is open the electrons stop their movement.

Translucent A material that allows some light to pass through, but changes the direction of the light.

Transparent A material that allows light to pass straight through.

Transverse Wave Waves that move up and down like water waves.

Weight The amount of pull that gravity has on an object.

Work Accomplished when a force is applied to an object and the object moves.

236

Thrust The force that moves an object forward.

Vibration The back-and-forth movement of materials.

Wedge Any material tapered to a thin, pie-shaped edge, such as nails, an ax, and a pencil point.

Index

239